T0303638

THE VISITATION

The earthquakes of 1848 and the destruction of Wellington

RODNEY H. GRAPES

VICTORIA UNIVERSITY PRESS

TE WHARE WĀNANGA O TE ŪPOKO O TE IKA A MĀUI

VICTORIA
UNIVERSITY OF WELLINGTON

VICTORIA UNIVERSITY PRESS
Victoria University of Wellington
PO Box 600 Wellington
vuw.ac.nz/vup

National Library of New Zealand Cataloguing-in-Publication Data
Grapes, R. H.
The visitation : the earthquakes of 1848 and the destruction of
Wellington / Rodney H. Grapes.
Includes bibliographical references and index.
ISBN 978-0-86473-686-4
1. Earthquakes—New Zealand—Wellington—History—19th century.
2. Wellington (N.Z.)—History—19th century. I. Title.
551.22099363—dc 22

Printed by 1010 Printing International Ltd, China

THE VISITATION

CONTENTS

Acknowledgements   7

Preface   9

Introduction   11

1.   Wrecked   15

2.   A most painful duty   23

3.   More of the same – only worse   31

4.   The uncertainty continues   49

5.   North and east   57

6.   The Middle Island   67

7.   Rap on the knuckles   82

8.   Generosity declined with thanks   91

9.   A nine days' wonder   100

10. Fire, water, air: A fissure in the ground   111

11. Interlude: Searching for an earthquake and an important discovery 124

12. The Awatere Fissure – source of the 1848 earthquakes   139

13. Future visitation   151

References   165

Appendix 1: Memorial and Schedule of Losses by the Earthquake   175

Appendix 2: New Zealand Earthquake Felt Intensity Scale   183

Index   185

ACKNOWLEDGEMENTS

The book would not have been possible without the research work on the 1848 and 1855 earthquakes done in collaboration with my co-awardee of a Royal Society of New Zealand 2002 Science and Technology Medal, Gaye Downes, formerly at the Institute of Geological and Nuclear Sciences. When I departed from New Zealand in 2001, Gaye remained as the sole intuitive gatherer, checker, integrator and interpreter of information gleaned from the remotest shelves of archival repositories concerning New Zealand historic earthquakes. Much of the information used in this book was collected by both of us as part of a New Zealand Earthquake Commission-funded project on the 1848 earthquakes and was supplemented by accounts previously obtained by George Eiby of the Seismological Observatory in Wellington and published in his 1980 Department of Scientific and Industrial Research Bulletin No. 225, *The 1848 Marlborough Earthquakes*.[1] The work of gathering and transcribing all the handwritten and printed material available on the 1848 earthquakes is published in a report entitled *Historical documents relating to the 1848 Marlborough earthquakes, New Zealand*.[2]

I extend my, or rather our, sincere thanks to the staff of the Alexander Turnbull Library, the National Library of New Zealand, the National Archives and the Wellington Public Library for all their help. I would also like to thank Alan Mason, Convenor of the Historical Studies Group of the Geological Society of New Zealand, for supplying copies of various documents relating to the 1848 earthquakes, and Eamon Bolger, archivist at the Museum of New Zealand Te Papa Tongarewa, who provided copies of Alexander McKay's unpublished manuscripts relating to geological investigations in the Awatere Valley, including his account of the 1888 Amuri earthquake.

Away from computer, office and library, I thank my former colleague, Professor Tim Little of the School of Geography, Environment and Earth Sciences, Victoria University of New Zealand, Wellington, for sharing his knowledge, intuition and companionship during field investigations of the Awatere Fault in Marlborough. Tim's work, together with that of his research students and his colleague Russ Van Dissen of the Institute of Geological and Nuclear Sciences, have put the Awatere Fault, and its movements during the 1848 earthquake and previous earthquakes, firmly within the international scientific arena.

My special thanks to Kyleigh Hodson of Victoria University Press for her expert attention and guidance during preparation of the manuscript for publication.

To my partner, Kyongran, I owe more gratitude than I can express in words.

Last and least, I want to take the liberty of crediting myself – for persistence in adversity.

PREFACE

Living most of my life in Wellington, New Zealand, I have grown up with earthquakes. Fortunately, most of them have been small and over before I had time to react, but now and then an occasional larger shake does cause concern, such as the first earthquake experienced by the early settlers in Wellington on 26 May 1840, two months after the township was founded, when they 'were aroused by an undulatory motion of the earth, and a somewhat severe shaking of their houses. Everyone seems immediately to have had suggested to their minds that it was an earthquake.'[1] As a child, I was taught that when an earthquake occurred I should quickly get under a table or a bed, or stand in a doorway, in case anything should fall. There were also periodic earthquake drills at school, where we were instructed to squeeze under our desks and then, when the 'all clear' signal was given by the ringing of the school bell, to file in an orderly manner out of the classroom into the middle of the playground.

I remember when the great magnitude-8.9 Chilean earthquake of 22 May 1960 occurred; I was at high school, situated on a low isthmus that joins the suburb of Miramar to the rest of Wellington. At lunchtime a special assembly was called and we were told by the headmaster that a tsunami had been generated by the earthquake and was racing towards New Zealand across the Pacific Ocean (I distinctly remember the word 'racing'). We were told that the town of Hilo on the main island of Hawaii, some 10,000 kilometres from the Chilean coast, had already been devastated by the arrival of the tsunami 15 hours after the quake, and as our school was sited not far above sea level we were instructed to go home for the afternoon. Great! An unexpected holiday! And with that many of us promptly cycled down to the wide, sandy beach at Lyall Bay to wait for the incoming wave.

After milling around for a couple of hours and seeing nothing, the expectation and excitement turned to boredom and we went home. The tsunami generated by that huge earthquake eventually did arrive later that evening, but it caused only a small rise in the tide.

So, with my life experience of earthquakes so far, I have become somewhat complacent and tend to remain sitting, standing or lying in bed rather than diving for the nearest doorway. I can only suppose that I'm waiting for the shaking to intensify enough to get me moving, but it invariably stops before I become activated. My complacency is a symptom of the knowledge I have gained from having studied, read and written about the effects of earthquakes, mapped earthquake fault lines and listened to numerous experts talk about them. I'm acting as if I know what is going to happen. But the reality is, I don't. Like everyone else, I cannot predict how an earthquake is going to develop. I imagine that when a large earthquake does occur I will react on the spur of the moment and instinctively, together with everyone else.

Of course we can minimise the effects of a large earthquake – fix furniture to the wall, anchor moveable objects with Blu-tack, lay in emergency food supplies and other necessities, as we are advised to do in the Civil Emergency section of the telephone directory – but we cannot do much about the psychological effects until afterwards. We live in an earthquake-active country and must learn to deal with it as best we can. Every major earthquake improves our preparedness – we attempt to learn from the past. But no amount of planning can really prepare us for the experience of a large earthquake. Individual experiences will differ depending upon each person's psyche just as much as where they are and what they are doing at the time. This I know from the personal accounts I have read, and it reminds me of the man who, when pulled from the rubble of the 1989 earthquake at Yerevan, Armenia, by a French rescue team, and thinking that World War III had begun, raised his hands and said 'I surrender'.

Rodney Grapes
Wellington,
December 2010

INTRODUCTION

It has often been said that the dread of earthquakes is strongest in the minds of those who have experienced them most frequently. In contrast to almost any other natural hazard, where familiarity makes people bold and more determined, no forethought can guard against large earthquakes because they may occur at any time, without giving the slightest warning. And when an earthquake has begun, no skill, presence of mind or preparedness can dictate the way to safety, because it is everywhere. For a comparatively short time confusion reigns. Although religious sentiments and the wrath of God are often invoked by these awful visitations, a sense of helplessness and a realisation of the futility of every human exertion against such a force all contribute to a general state of anxiety and panic.

The effects of earthquakes on the earth's surface, and on anything that grows from, covers, sits, stands or is built on that surface, have been repeatedly documented since history began. Along with this, a host of superficial phenomena relating to earthquakes have also been recorded: irregularities in the seasons before or after, sudden gusts of wind interrupted by calms, heavy rain at unusual times, the release of electricity or gas from the earth, the smell of sulphur, strange atmospheric effects and appearances that have both amazed and terrified the observers, underground noises that sound like the movement of heavy vehicles or railway engines, animals uttering cries of distress, and sensations like sea sickness or dizziness.

Such were the experiences of the inhabitants of Wellington during a cluster of devastating earthquakes that began on 16 October 1848. The fearful violence of the shocks, the destruction of property, and the frequency and long continuance of the danger caused universal alarm and produced an undefined sensation of terror. So exclusive was the concern for life that the loss of property was scarcely thought of, and the first reaction of many was to leave the uncertainty and misery behind if they had the means and opportunity to do so. During the height of the earthquakes, a feeling of despondency entered the minds of many, and it was feared the settlement and its future were ruined. But Wellington did recover, quite quickly. The British immigrants were not going to be deterred after coming halfway around the world and facing the trials and tribulations of a four-month-long sea voyage. In any case, most had nothing to go home to. Fortitude and stubbornness began to take root. Amidst the wreckage of houses and ruined possessions, 'the finger of God' was acknowledged – 'It is the Lord's doing, and is marvellous in our eyes'. The earthquakes had certainly been a calamity, an unforeseen happening against which no prudence could guard, and to which the inhabitants of Wellington, being God-fearing folk, submitted with due resignation. They believed they had deep cause for thankfulness that amidst all the chaos and confusion, occurring first in the dead of night, and later in the light of day, 'the visitation' had resulted in only three deaths.

Would it be any different today if the same sequence of earthquakes were to occur? The answer has to be, emphatically, yes. With a population of some 300,000, compared with just over 3000 in 1848, 'the wreckage of houses and ruined processions' in Wellington would be immeasurably greater during a similar series of large earthquakes and, depending on the hour of the day, the number of potential casualties would also be much higher. The cost would be enormously greater, as would insurance payouts, but the terror, anxiety and confusion would probably be the same, although there would be few who would believe that the earthquakes had been sent as a result of the wrath of God. The ground on which everything stands remains the same except for the reclaimed extension of the city into Lambton Harbour. And therein lies the unknown, despite all the predictions of what could or would be expected and the precautions implemented. Putting aside the sophisticated earthquake engineering know-how that has been developed in New Zealand in recent years to contend with the shaking effects of future large earthquakes on buildings, one can only imagine the experience, and learn from accounts of those who have been through such a devastating event.

This book, like an earlier one, *Magnitude Eight Plus*, which describes the effects of New Zealand's biggest historic earthquake on 23 January 1855,[1] is based on such experience – first-hand accounts in letters, diaries, newspapers and Government papers. It is also an account of social response to an unpredictable and heart-rending event where one is powerless to do anything but wait until it is ended. For many early Wellington settlers, earthquakes were part of living there. Since the founding of the settlement in 1840, there had been a couple of reasonably large shakes before the devastation of 1848, followed by the 1855 mega-quake that again wrecked the town. With two big earthquakes in seven years, many reminiscences written between twenty and thirty years later got them confused. And if they were confused then, they are forgotten now. Perhaps it's time to remind ourselves of past events that could in future reshape our lives and attitudes. Indeed, at the time of completing this book, the big one did arrive – a 7.1 magnitude earthquake at 4.36 on the morning of Saturday 4 September 2010, wreaking havoc in Christchurch. It was totally unexpected (everyone thought that Wellington would be the most likely next target) and was caused by movement of up to 4 metres horizontal and 1 metre or less vertical along a completely unknown fault, or faults, buried beneath the river gravels of the Canterbury Plains over a distance of at least 22 kilometres. Luckily there were no fatalities, but damage to buildings, ground and possessions was extensive. The insurance bill will be enormous. Aftershocks are expected to continue for quite a while, probably many months, but with decreasing intensity and number, although one can never be sure.* With today's expertise, much will

*Since completion of this book, a further earthquake of magnitude 6.3 struck Christchurch at 12.51 pm on 22 February 2011. Tragically, 181 lives were lost and damage was much greater than that caused by the earlier earthquake. This was because it struck at lunchtime, when most people were up and about, and the shock was centred only 10 kilometres from the city centre and at a shallow depth of 5 kilometres, resulting in greater shaking intensity. Two hundred kilometres to the west of Christchurch, the earthquake dislodged an estimated 30 million tons of ice from the Tasman Glacier into a nearby lake, causing 3.5-metre-high waves.

be learned from this earthquake, but from the numerous reports still emerging, there are many similarities with what happened during the 1848 earthquakes that devastated the infant town of Wellington.

Along with describing the effects of the 1848 earthquakes and the social response to them, this book also explains the cause – the relationship between earthquakes and movement on fault lines, first discovered in the late 1880s by the celebrated New Zealand geologist, Alexander McKay, following a large earthquake in North Canterbury on 1 September 1888. It also discusses the fault that ruptured during the first great shock of the 1848 earthquakes, what remains of the 1848 rupture today and its relation to large earthquakes in the past, and the geological context of the 1848 earthquakes as products of the convergence of two great earth plates through the Marlborough–North Canterbury area of the South Island of New Zealand. The book ends with a possible scenario of a future large earthquake in Wellington.

CHAPTER 1

WRECKED

Wednesday 25 October 1848 dawned clear and sunny, a good day for the advertised sailing to Sydney – and none too soon after a harrowing and terrifying week quarantined in Lambton Harbour, fronting the town of Wellington. Preparations were well under way, and by mid-morning the last of the cargo was being stowed and the first of the sixty-odd passengers with their belongings were being ferried out in the ship's boats. In his cabin, Captain Mills read again with considerable satisfaction the testimonial presented to him the day before on behalf of the good people of Wellington.[1]

> Sir, – We the undersigned Inhabitants of Port Nicholson, duly appreciating the promptitude with which you offered your vessel, the *Subraon* of London, as a place of refuge to those amongst us, who might wish to avail ourselves of your extremely kind offer (during the calamity with which this settlement has been lately visited) and deeply impressed with the constant attentions and almost unequalled liberality shown by you towards those on board, are anxious to take this opportunity, before your departure, of expressing our warmest thanks for your benevolent conduct.
>
> We beg you will accept the accompanying purse to purchase a piece of plate, bearing the annexed description, as a trifling proof of our gratitude, and at the same time receive the assurance of our best wishes that (wherever may be your destination) you will continue to enjoy health and prosperity.
>
> We are, Sir,
> Your most obedient and grateful servants, etc.
> Major Baker J.P.
> W. B. Rhodes,
> Geo. Moore.

He read the enclosed inscription and pictured how it would look on the silver plate he planned to buy in Sydney:

> Presented to Captain J. P. Mills of the *Subraon* of London, by the Inhabitants of Port Nicholson, New Zealand, as a mark of their esteem and grateful acknowledgement of his attention to the settlers and passengers on board his vessel, during the serious calamity with which the Town of Wellington has been visited – October 23rd, 1848.[1]

Sailing was delayed until late in the afternoon when James Calder, pilot, 'purveyor of fish', and ferryman for those wanting to cross the harbour entrance, arrived from the pilot station on the south coast and came aboard.

The anchor was weighed and the *Subraon*, reputed to one of the fastest ships on the Australian run, headed slowly out into the harbour under a fresh southerly breeze. The deck was crowded with warmly dressed passengers, 66 men, women and children[2] who forlornly and silently lined the starboard side to catch a last glimpse of the town where their hopes and plans for a new life had been terrifyingly ended by the 'visitation'.

At anchor in Lambton Harbour, Wellington, 1846. The ship on the left is the 486-ton barque, *London*, a smaller version of the 510-ton *Subraon* wrecked on Barrett Reef at the entrance to Port Nicholson en route to Sydney, 25 October 1848. To the right is the Colonial Government vessel, HMS *Driver* (ATL A-138-162).

Wellington in 1851, by Charles Barraud (ATL G596). In the centre is the Te Aro part of the town; in the distance, beyond Clay Point, is Thorndon and Pipitea Point, and the ships are anchored in Lambton Harbour.

With its backdrop of wooded and partly cleared hills, Wellington presented a clean and agreeable but primitive appearance compared with an English provincial town. Eight years in the making, the principal street of the settlement – called 'the Beach' – ran along a narrow space between cliff and high-tide mark, and was lined with shops and stores. Inland, the flat areas of Te Aro and Thorndon lay at either end of the 'beach'. They were dotted with small white houses that looked like scattered sheep, with here and there a commodious store, a large dwelling house or a creditable public building. Those on board the *Subraon* were leaving behind a town of some 3250 people. Wellington was a centre of supply for the recently established settlements of Nelson (established in 1842 and an average of three days' sailing distance from Wellington), New Plymouth (1841, two days' sailing) and Wanganui (1840, two days' sailing); it was also a depot for troops, and enjoyed, together with Auckland, a monopoly of government expenditure and substantial parliamentary grants. For these reasons, and because of a deficiency of open, arable land, Wellington had become a commercial centre, with most of its inhabitants engaged in trading pursuits and shopkeeping. But there was also a solid upper-class substratum of professional people – doctors, lawyers, civil engineers, architects, administrators and Provincial Council members – some of whom owned large quantities of sheep and cattle in the Wairarapa and Rangitikei areas and could be regarded as landed gentry. The population's spiritual needs were served by an abundance of clergy – two Anglican ministers, a Roman Catholic church with a bishop and several priests, a Wesleyan minister and a large chapel, a Scotch church, Independents, Primitive Methodists, and one or two of other denominations – and an influential Jewish community. All had had a busy time spiritually consoling the populace over the last few anxious days.

The *Subraon* sailed out of Lambton Harbour past Point Jerningham (named after Jerningham Wakefield, a prominent member of the New Zealand Company, which had founded the Wellington settlement), which formed the eastern headland of the harbour. Crossing the wide mouth of Evans Bay, Captain Mills cast a wary eye south across the low isthmus of sandhills at the head of the bay and noted that clouds were building. As they passed Point Halswell, the eastern headland of Evans Bay, Wellington was finally lost from sight and the ship turned south for her final run out into Cook Strait. It was slow going as she tacked towards the harbour exit, 'the Heads' as it was known, against a southerly swell and a freshening wind.

The task was made more difficult by the narrowness of the harbour entrance, bordered to the east by steep, barren hills and rocky coast, and to the west by a broken finger of black rocks known as Barrett Reef. In a strong wind, either from the northwest or the southeast, as in this instance, sailing ships were frequently unable to get through this narrow passage for several days. So, with the light fading, the wind and swell rising, and only a few hundred yards to clear the harbour entrance, Pilot Calder, against the advice of Captain Mills, issued instructions to turn the ship to starboard and take her through a narrow gap in the reef called Chaffers Passage.

It was a risky manoeuvre, but the pilot had to be picked up by a boat sent out to meet the ship after she had cleared the entrance. His house was in Tarakena Bay, a small cove just west of the harbour entrance, and Chaffers Passage meant a shorter distance and less strenuous effort for the boat rowers to reach the ship. Under port

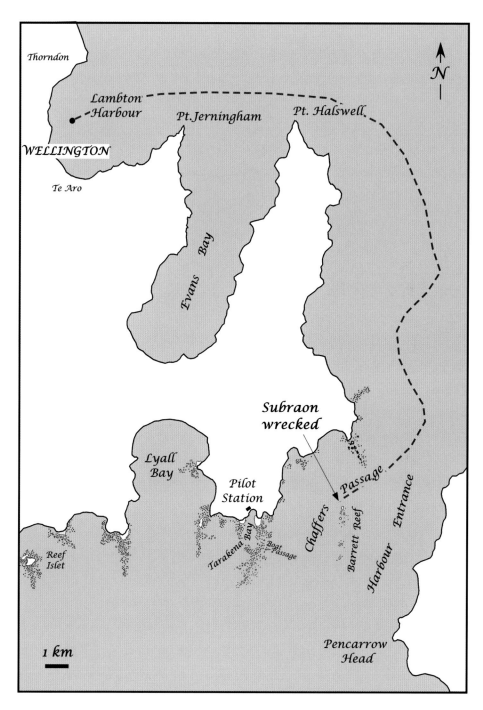

Map showing the location of Barrett Reef, Chaffers Passage, where the *Subraon*, carrying those fleeing the earthquakes, struck a rock and foundered on the evening of 25 October 1848, and the Pilot Station, where many of the passengers sought refuge.

regulations, Calder was in charge of the ship until she had cleared the Heads, and he had no intention of finding himself on an unscheduled voyage to Sydney because of adverse weather conditions preventing him from disembarking after completing his duty.

The ship had almost cleared the passage and was about to make a last tack when she ran out of time and space. The pilot attempted to turn the ship away from the wind but there was insufficient room, and at 8 o'clock a sickening sound of splintering timber signalled that the 510-ton barque had struck rocks on her port side.

In a heavy swell, with the light almost gone, Calder had grounded the *Subraon* in a perilous position – but fortunately only a hundred yards from the shore and three quarters of a mile from the pilot house. The passengers were confused and panicking. The rudder had gone and the hold was rapidly filling with water. Captain Mills immediately resumed control, and ordered his officers and crew to make preparations to land. Two of the ship's boats and the boat that arrived to retrieve the pilot were filled. The women and children were taken off first and, with some difficulty and danger, they managed to reach a small bay a mile east from the pilot house. Many of the women were without shoes, but they made sandals from pieces of blanket and clothing, and so were able to trek along the rocky shore to the shelter of the pilot house. Later that evening a few hardy individuals walked the ten or so miles back to Wellington in the dark, where they arrived at 3 o'clock in the morning to raise the alarm.

Early next morning the weather had calmed and several vessels left Wellington at an early hour to help save the cargo and the property of the unfortunate passengers. Captain Oliver of the 18-gun HMS *Fly* sent off five boats with a large contingent of men. George Moore, one of those who had presented Captain Mills with the testimonial, sent his schooner, *Gypsy*; and the agents of the *Subraon*, Messrs James Smith & Co, dispatched several more boats with about thirty men. All that day they were kept busy landing the cargo. It was hoped that the fine weather would continue until they could refloat the vessel and bring her around into the relative safety of the harbour.

Many of the townspeople ventured out to see the wreck and lend what support they could. Captain Charles Ewen and his fellow officers of the 65th Regiment were also determined to go, but as four men were to be flogged with 50 lashes each at parade in the morning, they were unable to start until the afternoon.[3] On their arrival, 'a melancholy spectacle presented itself', wrote Ewen. The *Subraon* was rolling to and fro on the rock with her nose in the water, and most of the passengers with remnants of their baggage were still encamped on the rough beach, sheltering under temporary tents made of sails, 'and all in a woeful plight'. By this time, hopes of refloating the ship had faded and it was expected that at high tide she would break up, with the loss of everything that remained. The passengers' lives had been 'mercifully spared', but many had lost what little they had taken on board with them, and by the end of Saturday they had nearly all arrived back in Wellington, much 'discomforted and dispirited'.

On hearing news of the wreck, Lieutenant-Governor John Eyre immediately suspended the pilot pending an inquiry. Captain Sharpe, the harbour master, was instructed to take over the pilot's duties and 'to look out for and board all strange vessels' in the interim. The Board of Inquiry, made up of a Mr Rhodes, Mr Luke (master of the *Fly*) and Sharpe, concluded that Calder, 'having placed the vessel in a perilous position . . . seems to have lost all presence of mind', and he was dismissed forthwith.[4] The *Subraon* proved to be unsalvageable, and after spirited competition the hull, masts, sails and anchors were sold for 515 pounds. For the unlucky Captain Mills, 'late of the *Subraon*', there was another letter of appreciation, this time from the passengers:

3rd November, 1848.

To Captain Mills, Commander of the barque Subraon,

We, the undersigned passengers of the barque Subraon, beg to offer you our most grateful acknowledgement for your conduct on the occasion of the wreck of that vessel, on the evening of the 26th inst., when under the charge of the pilot. To the presence of mind and intrepidity, to the calm courage, and the exercise of those qualities which distinguish a British sailor, displayed by you and those under your command, on that trying occasion, we attribute, under God's providence, the preservation of our lives, and those of our families, and the deliverance from the imminent danger to which we were exposed.

Begging you will accept our best wishes for your future prosperity,

We remain,
Your grateful and obedient servants.

William Fitzherbert, J.P., Sarah Jane Fitzherbert, Robert Park for family, S. Wilson, May and Emma Hartley, J. Walden and wife, Jabez Dean for wife, Richard Beamish, W. P. Young, Captain 65th regt., William Spinks for self and family, J. Hill, John Brown for self and family, F. Bishop and family, W. Smith and family, John Harvey, J. T. Hansard, M.D., and family.[5]

The last mentioned, Dr James Hansard, had lost all his baggage, but fortunately a box was picked up that contained his money so that he was able to sail for Sydney soon after.[6] Also lost were letters and newspapers sent to England by Captain Ewen. Most of the other passengers reluctantly decided to stay after their ordeal at sea and to start rebuilding their lives and houses that had been so suddenly shattered by the 'visitation' ten days before.

CHAPTER 2

A MOST PAINFUL DUTY

'It is my most painful duty to inform your Excellency that a terrible calamity has overtaken this province. An earthquake has occurred, and the town of Wellington is in ruins.' These direct and ominous words were penned by John Eyre, Lieutenant-Governor of the Province of New Munster, at the beginning of a communication to Sir George Grey, Governor-in-Chief in Auckland, on 19 October 1848. The report went on to describe the effects of a cluster of violent earthquakes that Eyre considered to have 'struck a blow at the prosperity, almost at the very existence' of the fledgling settlement.[1]

At twenty minutes to two in the morning of Monday 16 October, the approach of an earthquake was signalled by a distant hollow sound travelling rapidly from the north and south. In an instant the town of Wellington was undergoing a violent and disastrous shaking that continued for nearly a minute; the motion being horizontal and undulatory until towards the end, when it seemed to have an upheaving or vertical movement.

After a relatively quiet interval of half an hour, another shock less intense than the first occurred, and during the following 19 minutes there was a succession of severe shocks, with lesser ones at intervals. During the whole time, the ground was in a state of constant vibration, and the fury of a strong southeasterly gale added its force to the destruction that was occurring on the ground.

For Captain Charles Ewen of the 65th Regiment, it was a truly 'horrible night'. Having been jerked out of bed by the earthquake, he fled naked into the yard, hardly able to 'keep his legs'. Looking back he distinctly saw the house rocking to and fro and the earth heaving 'convulsingly'. Part of the brick gable of the house collapsed in a

Sketches by Captain Thomas Collinson in 1849 of Wellington. *Above*: View from Thorndon looking into the garden of Government House, with its flag and signalling mast. Mount Victoria is in the background and Collinson's ship HMS *Fly*, on which he felt the earthquake at 1.40 am on 16 October, lies at anchor second ship from left. (A-292-070). *Opposite*: View from the same Thorndon location towards Te Aro (in background), with the Maori Pa (on beach front) and military barracks on Mount Cook behind. In the centre are Lambton Quay and Clay Point (ATL A-292-071).

cloud of dust just as the landlord's family, who were sleeping in the upper storey of the lodging house, rushed downstairs and out of the building. As soon as the agitation had diminished, an embarrassed Ewen ventured back to his room to get dressed, which he did hurriedly, fearing that at any moment the walls were going to collapse.[2]

A few hours after Rugby Pratt had gone to bed 'amid a hurricane of wind and an almost ceaseless downpour of rain' that had already flattened fences and flooded various parts of the town, he was suddenly awakened by an unusual vibration that was clearly something other than the effect of the wind. Sleeping in the upper part of a store adjoining a large building, he was uncommonly nimble in leaping out of bed, and getting to and opening the outer door. Not caring to face the wind and rain in his light nightdress, he remained just within the shelter of the passage, ready for an energetic leap when the expected crash came. After eight or ten minutes the crash

had still not happened, and as the successive shocks did not have same violence he felt calmer and went back to bed, though he was still very wary as to the safety of the building. There he remained, wide awake because of the perpetual motion and numerous 'smart' smaller shocks, until daybreak.[3]

Nearby, in their small wooden house, Thomas King and his wife Mary had quickly dressed after being woken by the shock. Not daring to remain in bed, they decided to spend the rest of the night, with their daughter Polly, by the kitchen fire. Unbeknown to them it was a dangerous place to sit; in the morning King discovered that from the roof upwards the chimney had been cracked. From the time of the first shock until eight in the morning they counted 42 distinct shocks.[4]

Mr Justice Henry Chapman, Judge of the Supreme Court for the southern division of New Zealand, in his new two-storey house 'Homewood' set in five acres carved out of the bush amongst the Karori hills above Wellington (where he lived with his wife, Kate, their four children, Harry, Charles, Martin and Ernest, a cook, two nurses and an outdoor man), wrote that the first great shock:

> did not cease its vibrations for 10 minutes and was very strong for about 2 minutes. The house rocked very violently. My first impulse was to gather up the

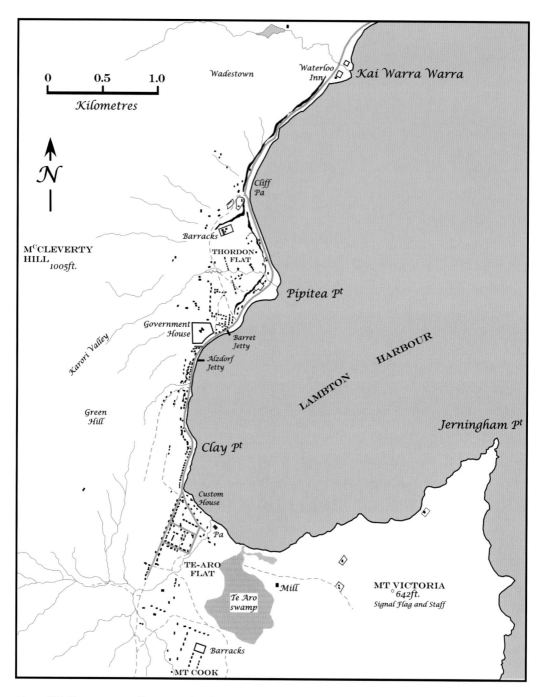

Map of Wellington town (from British Admiralty hydrographic map): Port Nicholson, 1856, surveyed by HMS *Acheron* in 1849. The thick brown line is the principal road, which mainly follows the shoreline; dashed lines represent formed tracks; black squares are buildings; enclosed areas are compounds that are fenced or stockaded.

younger children rouse Harry and rush out but a moment's thought produced a conviction that nothing would go but the chimneys and they were not in a position to hurt anyone. For an hour the vibration was excessive – every few minutes increasing to a shock. By 6 in the morning I should think about 100 shocks must have occurred but they seemed to diminish in frequency and in force.[5]

The house was fortunately built of wood and survived the shaking except for cracked chimneys. Internal damage was minimal (the servants' bells were set in motion and the clocks stopped), despite the fact that for the first three quarters of a minute Chapman could barely stand.[6]

Roused from bed by the first shock, Lieutenant-Governor Eyre and his reluctant staff ventured out into the wild night with a lantern, rather curiously to look for cracks in the ground – the Lieutenant-Governor having read that such features often formed during earthquakes, and that they sometimes opened and closed during successive shocks. He soon discovered a good number in the soft soil but was disappointed not to find any of great length or depth. Arriving at Colonel Gold's house nearby he was amazed to find that the officer commanding the troops of the 65th Regiment stationed in Wellington 'did not feel it part of his duty to get out of bed to meet such an enemy'. Annoyed at being woken by Eyre at such an early hour, the Colonel grumbled that he had 'shaken his bed more than any earthquake did'.[7]

Colonel Charles Gold, Commander of the 65th Regiment in Wellington at the time of the 1848 earthquakes (ATL F71999 1/2). He was not amused at being woken by Lieutenant-Governor John Eyre soon after the first great shock at 1.40 am on Monday morning.

'Homewood', the house of Judge Chapman, Karori, near Wellington (ATL F 21070-1/2). The first of the 1848 earthquakes early on Monday morning, 16 October, cracked the main chimney. The third great shock, early in the morning of Thursday 19 October, destroyed all the chimneys, caused considerable breakage in the storeroom and demolished the plaster in two sitting rooms.

Mr Justice Henry Samuel Chapman, Judge of the Supreme Court for the southern division of New Zealand and a methodical recorder of the effects of the 1848 earthquakes in Wellington (ATL B-039-101).

Ships in the harbour had also felt the effects, the sensation experienced on board HMS *Fly* suggesting that the vessel had suddenly grounded. Captain Oliver was awakened by the violent motion, and the sentry running into his cabin to answer the bell that 'had not been rung by me'.[8] The shock was felt most in the forepart of the ship and all the men, including Oliver, ran up on deck as soon they could. Elsewhere, the earthquake had been noticed on ships at sea. About twenty miles off Cape Campbell in the northeast part of the 'Middle Island', the *Sarah Anne* felt the shock 'very heavily'. The ship shivered from stem to stern and the captain and crew thought she had struck a rock and was forging over it. A lead was immediately thrown overboard, but no soundings were found. The *Clara*, en route to Wellington from Auckland, also experienced the shock sixty miles out from the coast.

At daybreak the 'scene of desolation and ruin that presented itself was a melancholy one – everyone, whole families deprived of their comfortable dwellings'.[2] But miraculously no one had been hurt. Several brick and solid clay buildings at Te Aro and in other parts of the town had been seriously damaged. Among those that had suffered most was the brick-and-cement Wesleyan Church. The impressive front of the building consisted of four pilasters surmounted by an entablature and pediment, with a central doorway, and a large window on either side. During the shaking all this heavy adornment acted as a lever, causing the front wall of the church to separate from the roof, lean outwards and overhang the road. Most of the chimneys in the town had been thrown down or were badly cracked and twisted. Reverend Samuel Ironside noted that 'it was pitiable to walk around and see nearly every brick chimney that stood above the ridge board of the house cut off from the ridge as cleanly as though done with a sharp tool, and slewed half around'.[9]

Destruction of glassware and crockery was considerable. In his storeroom, Rugby Pratt and his companions found that the shelves had been cleared of nearly all their bottles, and the floor covered with the debris of what had been bottled fruits, pickles, and salad oil. They were amazed to see that a large tank, recently filled with about two hogsheads of whale oil, had been hurled from its usual position to the opposite side of the store, losing half the contents during its flight. Throwing down bags of sawdust, and working vigorously with barrows and shovels, they had by midday cleared away the mess. With something like order once more restored, they noticed the dangerous state of one of the chimneys. The portion that projected above the roof – about six feet – had been broken off near the roof, and turned round on the stack without being thrown down, as observed by Reverend Ironside. From this, Pratt conjectured that the earthquake's motion must have been rotary as well as undulating.

'Some of the freaks of the giant power on its way through the town were marvellous,' wrote Reverend Ironside. One such 'marvel' occurred at Te Aro, behind the damaged Wesleyan Church where the buildings were close together. Here, on Manners Street, stood Dr James Hansard's brick house, which contained stores of drugs and shelves of bottles along the north wall of one of the rooms. Next to his house, on its northern side and barely two feet away, was a wooden shop full of crockery, the south wall of which was lined with shelves of popular delftware. In the doctor's house the earthquake had smashed all the medicine bottles, whereas in the crockery shop next door scarcely a plate, dish, or cup had been moved.[9]

At Karori, Judge Chapman thought that most of the shocks had come from approximately north by east or north-northeast, with one particular shock seeming 'to have a double source, meeting about this neighbourhood, The twisted appearance of some of the chimnies confirms this'.[6]

All that day and into the night slight tremors of the earth continued at short intervals, each preceded by a low but unmistakable rumbling sound. Wellington's inhabitants remained in a state of 'greatest excitement', dreading a second shock. That evening, as the weather cleared and stars made their appearance, spirits began to revive, although few could sleep well during the night.

CHAPTER 3

MORE OF THE SAME
– ONLY WORSE

Tuesday 17 October

Tuesday was a remarkably fine day with a gentle wind from the southeast. Business in the town was almost at a standstill and it was the first day of the band season. In the afternoon, the band of the 65th regiment assembled on Thorndon Flat for the advertised promenade, which on this occasion was especially well attended. Chairs and camp stools were placed on the grass, with the band occupying a central position. The sight of the polished brass reflecting the sunlight and the bandsmen in their uniforms of snowy white picked out with red gave the occasion a festive atmosphere. The officers in dress uniform, and the fashionably attired wives, sisters, cousins, and aunts of the military, commercial and professional men of the settlement added to the brightness of the scene. The wife of the Hon. Henry Petrie, described as 'very young looking and with wild spirits, and enjoys a ball, a ride, or a scamper of any kind, and is sometimes very pretty',[1] had come from the Hutt Valley in her gig for the first time since her confinement, and was the life of the gathering. One would never have guessed that Wellington was still feeling the effects of a violent earthquake in the early hours of the previous day. The young settlement was putting on a carefree face.

A number of people commented on the 'unnatural glow in the air'. William Swainson, the Attorney General and noted naturalist, and several others were 'chatting merrily' with Captain Oliver of HMS *Fly* when they distinctly felt a slight shock. Oliver raised his hand and said 'There!' This was followed by laughter, with

Mrs Petrie pointing and saying, 'And don't you see that chimney falling over there?'
At that very instant a rumbling noise like a 'pack of artillery galloping over a bridge'
suddenly overwhelmed them. It was a noise that seemed to occupy all the space
above, below and around, so that one could almost feel and breathe it, and it was
almost immediately followed by a rolling motion that made the earth on which they
were standing or sitting look like a carpet 'being shook'.[2] Everybody had difficulty
standing, and many were thrown to the ground, along with all the chimneys in sight
– including the very one that Mrs Petrie had pointed out in jest 'and from which
she had not taken her eyes or finger for all this passed in much less time than it can
be said'.[2] The line of soldiers standing to attention for an inspection of arms rose
and fell as if a wave passed beneath them. The band stopped short as 'musicians,
music-stands, instruments and fluttering scores became inextricably tangled by the
unprogrammed tremolo movement'.[3] Everybody left the scene in anything but a
leisurely manner.

Judge Chapman was lunching with the Lieutenant-Governor at the time
and noted later that 'a loud sound warned us of a strong shock – the sound was
immediately followed by the violent agitation of the house'. The French windows
burst open, 'bottles flew from the table – books from a side table and two valuable
chronometers on the mantle shelf were thrown down. I picked them up and one
had stopped. The Lieut. Governor less accustomed to earthquakes than myself
rushed out of the house and called me to follow. When I went out all the servants
had assembled alarmed on the lawn. The vibration of the house had not ceased and
the cries along the beach warned us that some serious damage had been sustained.'[4]
Having an appointment at his Chambers at 4 pm, Chapman found that the 'short
stout chimney' of the Court House had collapsed. Looking towards the business
part of town, Te Aro, with his telescope, he could see that most of the brick buildings
had sustained damage: 'Chimneys lay prostrate in all directions. They are generally
built outside, and against the gables of the houses, so that happily no harm was
suffered by the inhabitants of the wooden houses.'[5]

Most people agreed that this shock was just as severe as the one on Monday
morning, but that it had been more disastrous in its effects, especially in the Te

The Courts of Justice, Thorndon, Wellington (ATL, PUBL-0020-19-2). It was here that Judge Chapman had his chambers, which he visited on Tuesday afternoon, 17 October, following the second big shock of the earthquakes. He found that the chimney between his room and the Registrar had collapsed and 'while there – the people with their spy-glasses discovered that several brick stores and the Methodist [Wesleyan] Chapel at Te Aro [in the distance] or south end [of Wellington] were partially down.'[4]

Aro and Thorndon areas of the town where very few of the already weakened brick and clay buildings escaped its destructive effects. As Chapman had observed, all the brick stores at Te Aro sustained further damage, their walls either collapsing or being cracked in different directions and thrust out of perpendicular. The Colonial Hospital in Thorndon had been kept from falling by its bonded timber framework, but was in a very damaged condition; the walls were cracked and strained in all directions, making it untenable. Patients were hurriedly removed to Government House. Those in the damaged Military Hospital at Te Aro were taken to the wooden barracks at Mount Cook, and prisoners from the gaol, whose walls were rent from top to bottom, had to be placed in the custody of the military. A large building in Thorndon that was used as a barracks for the soldiers was so

The Barracks, Thorndon, Wellington, by Samuel Brees, 1847 (ATL F81901 1/2). The building was reported to be so severely cracked by the first great shock that it had to be vacated. Inexplicably, the damage was not listed in the Schedule of Buildings Damaged that was sent to the British Government and published in July 1849.

John Plimmer (ATL 1/1-018604-F). He described his predicament at the top of a long ladder trying to repair Mr Rhodes' bonded store at the time of the second great shock of the 1848 earthquakes on the afternoon of Tuesday 17 October as 'horrid and perilous'.[6]

fractured that it also had to be vacated. Strangely, buildings along Lambton Quay and those situated on higher ground did not suffer so severely, except for their chimneys, which were almost without exception thrown down or more-or-less twisted and cracked.

The afternoon shock could not have come at a worse time for the builder, Mr John Plimmer. The early Monday morning shock had caused some damage to the large brick stores at Te Aro, especially those with heavy slate roofs, by breaking the bond of the brickwork and splitting the angles. One of these stores belonged to Mr Rhodes, who called on Plimmer for advice and help to prevent the end gables from falling. Plimmer thought it best to try and clamp them together with iron rods and plates. These were at once prepared, and bricklayers employed to make holes through the walls for their insertion, a task that they completed soon after three o'clock on Tuesday. Plimmer then climbed up two ladders lashed together to reach the gables. He had just reached the eaves where he planned to insert the tie rods, and had caught hold of the slates to help him lean over to get a clearer view, when the second great shock commenced. In such a 'horrid and perilous position', and thinking that he would go down with the building, Plimmer tightened his grip on the roof slates on either side of the ladder as the shaking intensified. The gable end fell with a tremendous crash. Sheer terror kept him hanging on until, after what seemed an eternity, the shaking slackened to a light quivering. Still paralysed with fear, he slowly recovered his breath and began to look about. Everywhere chimneys were down, houses seriously damaged, and he could see that the gable end of the Wesleyan Chapel in Manners Street had collapsed. The most curious thing that caught his eye was the way the Te Aro bog was still moving, 'shaking like a jelly . . . but more like a field of waving corn in a high wind'. He looked down to see how the ladder had been prevented from slipping sideways, and was astonished to discover that it had ground its sides through two thicknesses of slates into the wall plates of the building. He lost no time in descending to 'terra firma', which at that time did not seem firm at all.[6]

Rugby Pratt was just handing a customer a receipt for settling his account in his now orderly store when the shock began. It caused them both to rush outside, only to see the customer's dray and team of bullocks, which he had left standing in front of the store, careering down the street. While the customer chased after his runaway transport, Pratt returned to the desk where he had been writing to find the floor once again strewn with loaves of sugar, 'and the desk much indented by their falling upon it from the top shelves'. Although he had taken the precaution of securing all the bottles and goods on the lower shelves by passing wire along the front of them, he had made the mistake of thinking that the heavy goods on the top shelves would be safe in a further earthquake.[7]

At 3.40 pm, Captain Ewen was walking along the beach (Lambton Quay) to his lodgings when the shock struck. Everything was on the move all at once, almost immediately followed by the sound of brick walls and chimneys falling. Arriving at his own dwelling, which was chiefly constructed of brick, he stood aghast before a complete wreck. The family had barely made their escape for a second time. Eventually, Ewen managed to get a party of soldiers to clear away his property from the ruins. Everything was damaged – crockery broken, the furniture scratched and knocked about, and all his 'little nicknacks' covered with mortar. He was obliged to put it all in the street in a heap, and proceeded to pack it all as fast as he could, not being 'over particular about the proper places for each article'. Before nightfall, Ewen was fortunate enough to get 'a smart wooden house for shelter' where he put all his goods, taking a meal with some of his brother officers who had kindly offered their house as a 'refuge' for the destitute.[8]

In the study of the wooden mission house attached to the Wesleyan Chapel in Manners Street, Reverend Ironside's brother missionary and 'old Hoxton College chum' Mr Creed, from Waikouaiti, was in the middle of a meeting with one of the Fellowship classes at the time of the shock. The big eastern wall of the church, only some sixteen feet from the study, began to vibrate from side to side to an alarming degree. Outside, Reverend Ironside was consulting the builder and his men, who were taking down the large stones of the pediment in front of the church that had been damaged during the Monday morning shaking. Other men were perched on scaffolding. Ironside 'trembled for their lives, [his] heart was in [his] mouth, as they swayed to and fro with the building'. Luckily, the walls and roof maintained their position, so the 'dear women at worship were preserved, and the men on the scaffold were able to descend from their frightful position'. They had had a lucky escape. The chapel walls, although still precariously standing, were split in all directions, so extensively as to render the building a ruin.[9]

The Wesleyan Chapel and Mission House, Manners Street, Te Aro, Wellington, by Samuel Brees, 1847 (ATL A-109-022). The brick chapel, particularly the elaborate masonry fronting the building, was severely damaged during the first shock on Monday 16th, and the Reverend Ironside and 31 others were forced to take refuge in the adjacent weatherboard mission house, During the second big shock on Tuesday afternoon, the walls of the chapel were split in every direction, so as to completely destroy the building.

Then came the worst of news. James Lovell, Barrack Sergeant, in charge of the commissariat stores in a large brick building in Farish Street near the beach, had been looking after the Government property under his care at the time of the quake. His two children, a girl of four and a boy of eight, were with him. The shock was so sharp and sudden that, before they could escape, a brick wall surrounding the adjacent stores of Mr Fitzherbert fell, burying them under its debris. Others rushed to their aid, to find the girl dead, and the boy and his father badly injured. They were taken to the relatively undamaged wooden hospital on Mount Cook and were later visited by Captain Ewen, who found the boy barely alive and Lovell badly bruised with large pieces of flesh torn from left leg, his wife 'moaning piteously by his side'. Later that night the boy died. The father lingered on in great pain. 'It was enough to appall the stoutest heart', Ewen noted in his diary, 'but we must be thankful to God that such a little sacrifice of life has been made.'[8]

In the meantime, Reverend Ironside, his wife and several of their friends had been forced to take refuge from their wrecked homes with Mrs Watkin in the mission house next to the ruined Wesleyan Chapel. In all they were seven families and 32 people. Ironside felt a great responsibility. The feeding of this large number was not of much consequence at the time, for little cooking could be done with the chimney down, and in any case nobody felt they could do justice to a meal. When after a day or so their appetites had not returned, the neighbourhood physician became concerned. He directed Ironside to procure stimulants – an absolute requisite, he said – and to see that everyone took some occasionally to calm their nerves. Going to one of the stores and explaining the situation, Ironside was ushered by the proprietor into the shattered storeroom, a pathetic sight with scores of broken bottles lying about, their contents mixed together and covering the floor. 'You are quite welcome, Sir, to any – all – that you want. I shall make no charge.' He kindly sent his assistant up to the mission house with a dozen of the bottles that had so far escaped the wreck, and, under orders from the doctor, small portions of the stimulant were served every two hours.[9]

Returning to his home in Karori at six o'clock, Judge Chapman was disturbed to find that the lower parts of his chimneys had been further damaged by the afternoon shock: 'the library chimney cannot be used; the parlour chimney, which goes through the centre of the house, has been secured with boards and lashings, so has the kitchen chimney. A small chimney in the wing of the house, used as a day-nursery, appears to be uninjured, except the top.' Mr Parnel, his carpenter, who had been in the process of securing the parlour and kitchen chimneys, told him that 'the tremulous motion of the earth did not cease for eighteen minutes'.[5]

Reverend Samuel Ironside (ATL PACOLL-6075-10). He published reminiscences of his experience during the 1848 earthquakes 43 years later in *The New Zealand Methodist*. 'In view of the general calamity, the widespread feeling of danger yet imminent, there was a consensus opinion that the city should be called to fasting, prayer, and humiliation before God; recognizing the truth in the 100th Psalm, He can create, and He can destroy.'[9]

On the Tuesday evening a strange light was seen to the northeast. Along with the violence and duration of the 'shakings', and the constant rumbling sounds beneath the earth since the afternoon shock, this led to the belief that a volcano must have erupted.

Wednesday 18 October

Early that morning, the barque *Clara* from Auckland anchored in Lambton Harbour. While slowly working their way though the Heads, those on board had felt the Tuesday afternoon shock, with a sensation as if the ship had hit a rock. She had brought down the Auckland newspapers, which recorded that his Excellency the Governor-in-Chief, George Grey, had arrived in Auckland on 3 October, that HMS *Acheron* was still in Sydney, and that the Government steamer, *Dido*, would shortly be leaving for England; but for the Wellingtonians at this particular time they contained 'no news of any interest'. Captain Crow reported that during the voyage he did not observe any natural phenomena that could 'lead him to infer the active operation of earthquakes'; the only remarkable thing observed was a bright light between two and three o'clock that morning.

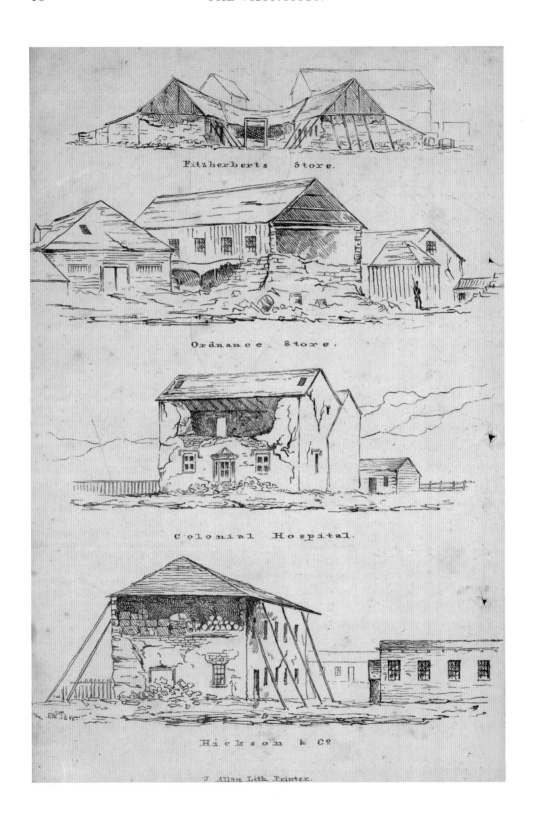

Fitzherberts Store.

Ordnance Store.

Colonial Hospital.

Hickson & Co.

J. Allan Lith. Printer.

Early in the day the tide had risen inexplicably to an unusual height, about 18 inches above normal high-tide level, overflowing part of Lambton Quay and all the area fronting the harbour at Te Aro, and flooding a number of houses and shops. All that day the shocks continued, at average intervals of ten minutes. They were usually preceded by a hollow rumbling noise, and in the background was a continuous trembling of the earth. Many people walked about, not daring to be inside or near any building, and a large number put up tents and awnings in open spaces. It was hoped that the light seen during the night was indeed from Tongariro or some other volcano in eruption, and that the pent-up gases in the earth were being released, lessening the chances of future earthquakes.

Lieutenant-Governor Eyre was conspicuous by his 'kindness and attention', visiting nearly all the shattered houses and consoling those who were seriously affected. Judge Chapman paid a visit to Te Aro to inspect the damage first-hand: 'It seems to have been the seat of the greatest force of the earthquake' with a large number of brick buildings 'rent to pieces. The walls just hold up the roofs, but large masses of brick-work have fallen out: all must be taken down.'[5]

Throughout the day, ministers of the several denominations held morning and evening services in those churches left standing and in some undamaged private

Facing page: Sketches of buildings damaged in Wellington by the 1848 earthquakes by Robert Park, surveyor of the New Zealand Company. They originally appeared in the *New South Wales Sporting and Literary Magazine and Racing Calendar*, 1848 (ATL, PUBL 0050-01). From top down: *Fitzherbert's store* (Manners Street, Te Aro), a single storey building, bricknogged with a brick boundary wall. The southeast and northwest ends collapsed together with the boundary wall. *The Ordnance store* (Manners Street, Te Aro), a two-storey building constructed of 13½-inch-thick brickwork with no wood bonding and a slate roof. The gables were thrown out and the north wall cracked in several places, although the side walls were little damaged. *The Colonial Hospital* (Pipitea Street, Thorndon), recently completed, was rendered unsafe by the second severe shock on Tuesday afternoon, 17 October. Patients were removed to Government House. It was a two-storey building with hollow brick walls 14 inches thick, with no bond timbers. The northeast gable and the front of the upper storey were thrown out on the south and east sides. The walls were cracked on all sides. *Hickson and Co's store* (Manners Street, Te Aro) had both gables thrown down. During the severe aftershock at 5 am on Thursday 19 October, the building was reported to have been 'thrown down' and the remaining walls rent in every direction. Evidently, Park's sketch was made before this happened.

houses. In the afternoon an inquest on the bodies of the two children of Sergeant Lovell was held at the Ship Hotel, before the coroner, Dr Fitzgerald, and a verdict of accidental death returned. The publication of the *Spectator* 'was unavoidably delayed till a late hour' because the greater part of the set-up type for the Wednesday issue had been thrown down and disarranged by the Tuesday afternoon shock. In the issue, Mr Lloyd, pastry cook, announced 'to the inhabitants of Wellington that, having repaired part of the damages occasioned by the late awful visitation of the Almighty', he would resume business as usual the following Saturday. By late afternoon the weather had changed to a cold southeasterly wind, making it uncomfortable for those sheltering under canvas. Darkening clouds betokened strong winds and rain to come.

That evening Captain Oliver and two of his officers from the *Fly* had an invitation to dine with Colonel and Mrs McCleverty, and although he expected a wetting, he 'was determined to go as people were beginning to put on long faces'.[2] After dinner they walked down to the beach to return to the ship, but the boat had not been sent, so they decided to seek lodgings for the night, his companions at Barrett's Hotel and Oliver at Baron von Alzdorf's Hotel on Lambton Quay. That night, large fires were lit outside for warmth and light, and those under temporary shelter gathered around them to spend the hours of darkness in 'singing, prayer and serious talk'.

Thursday 19 October

Except for a strong southeasterly gale and rain, the night passed uneventfully until, at a few minutes after 5 am, another violent shock accompanied by a loud roaring noise rocked Wellington. During the shaking, it was observed that the gale stopped as if 'bayed back', only to spring up again afterwards with the same fury as before. The first impulse of most people was to rush out and see the extent of the damage, many leaving their houses with good reason, and others from sheer fright.

Judge Chapman recorded that this shock was longer than the previous two.

> The extreme force of the shock lasted rather less than a minute, there was considerable motion for 3 ½ minutes, and the vibration lasted 8 minutes from the commencement of the shock. It has done more damage than all the others

together. It has split the solid bed of brickwork which forms the lower part of our oven, completed the destruction of other chimneys, torn the plaster of our lower rooms to pieces (the upper are lined with wood), and broken a great many loose articles. Our windows (French casements) flew open.[5]

Captain Oliver felt quite 'secure' in his low wooden building, having been previously told, 'which I afterwards found incorrect', that there were no heavy goods in the store above.

Directly I awoke I timed this shock and found from the time I took my watch till the shock ceased occupied 50 seconds, allowing 10 seconds for awaking and happening to note so, which perhaps is too much, would make it last a minute. In the next half hour we had 10 tremors or so, one every 3 minutes. After that I went to sleep again and did not feel anything severe until about ½ past 7 when I was shaving. I had to hold my hand several times, and moving about the passage in my dressing gown I found Bell and Domett who had been turned out of their houses sitting by the fire. Poor D. had been a good deal alarmed and I not knowing it said D. don't you know that that is a brick chimney you are sitting by, when up he started and ran out from the house.[2]

For those who went outside another 'melancholy scene presented itself to their view'. The Wesleyan Church, the brick Bonded Store of Messrs Rhodes and Co., the end walls of Messrs Ridgeway and Hickson's store, and the Ordnance Store, all lofty and substantial buildings, had finally collapsed; the end walls of the gaol were destroyed, and the boundary wall rent from top to bottom; the walls of the Colonial Hospital were broken and partly thrown down. Many other smaller brick and clay dwellings were 'levelled with the ground' and others, built of the same materials and not previously injured, were all now seriously damaged.

Shortly after the shock, the Lieutenant-Governor again went on a tour of inspection of what remained of the town, suggesting measures to assist and ensure the safety of the inhabitants, and offered the use of Government House to those in need of shelter. Walking along Lambton Quay, looking at the damage, Rugby Pratt passed Alzdorf's Hotel and was surprised to see a large number of 'thirsty souls' busy drinking so early in the morning. The breakage and loss during the week had been so great that the Baron thought he might as well distribute the balance of his stock gratis. His gesture was eagerly taken advantage of.[7]

At nine o'clock, the New Church at Te Aro and the Scotch Church were 'thrown open' as places of refuge for the homeless. Colonel Gold went to the barracks and ordered out fatigue parties of the 65th Regiment to give assistance. Under the direction of their officers the men spent all day clearing away the rubble of fallen buildings and warning people where danger threatened from unstable buildings still standing. Sentries were stationed to protect recovered property from plunder.

The latest shock had caused a great deal of further breakage in the shops. One tradesman had about £100 worth of sauces, pickles and preserves precipitated onto the floor from shelves that faced either north or south, whereas those sitting on shelves facing east or west, although moved around a little, had remained undamaged. A publican observed the hanging lamps of his rooms swinging in a southwest–northeast direction and that his weighty slate-based billiard table had been moved an inch from the wall. Near the Ordinance Store, just before it became a ruin, a barrel lying on its side suddenly jumped upright.

In the afternoon, news reached Wellington from the Hutt Valley that, before sunrise, a policeman and his companion standing on the bridge over the Hutt River had seen the river 'greatly agitated'. A few seconds afterwards, the bridge shook beneath them and they observed the road leading to the bridge rise and fall like 'the waves of the sea'. The motion gradually approached from the south, and the two men distinctly felt it pass under them.

> So awfully terrific was the shock on Thursday morning – so universal and great the destruction of property – so imminent and alarming the danger to life – so directly from the hand of God himself was the calamity, – and so utterly powerless was human aid to effect deliverance – that prayers and supplications to Heaven were the only means that could avail in this extremity. Accordingly the feeling that most extensively pervaded the community was that a day of fasting, prayer, and humiliation, ought to be observed.[10]

So thought the church leaders of the community and so thought his Excellency, the Lieutenant-Governor. Accordingly, on Thursday afternoon, a Government Gazette was quickly published, containing the following proclamation:

> Whereas it has pleased Almighty God to visit this settlement with a great and grievous calamity, it is fitting that a public acknowledgement be made of the divine power, on whom all the operations of nature and the security of his creatures depend, and that prayers and supplications be offered up to Almighty God to avert the recurrences of any similar visitation. Now therefore I, Edward John Eyre, Lieutenant-Governor of the Province of New Munster, by and with the advice of my Executive Council, do hereby proclaim and declare that tomorrow, the 20th of October, shall be held as a day of public fast, prayer, and humiliation.
>
> Given under my hand and seal, at Government House, Wellington, this 19th day of October, 1848.
>
> E. Eyre[11]

Notices were also published in the Gazette by the Reverend Cole, announcing that morning and evening services would be performed in the Episcopal Churches on Te Aro and Thorndon Flat, and by Father O'Reilly that the Holy Sacrifice of the Mass would be offered up at half-past ten. As the Wesleyan Primitive Methodist and Independent chapels had been destroyed, the Ministers of the Evangelical Alliance announced three services to be held in the Scotch Church. They also announced that services would be held at the Hutt, Karori and Johnsonville, and on board the *Subraon*, where a large number of families had taken refuge.

Thursday night, although calm, was punctuated by a great many tremors in quick succession. A fiery glare, widely regarded as a sign of the doom and destruction that had descended upon them, was seen among the clouds to the south from 8.30 pm, becoming gradually fainter until it disappeared at twelve o'clock. The lurid appearance of the sky, the continuing shaking, the constant rumbling sounds beneath the earth, and other indications reinforced the general belief that a volcano had almost certainly broken out in the mountains near the centre of the Middle Island,

and the hope was fondly cherished that if the dormant embers of some smouldering crater had been kindled up and burst forth – if some close up volcano was come again into a state of activity, a safety valve would be opened by means of which the pent up subterranean fires, that by their explosive force are shaking the earth in all directions around us, would be allowed to escape, and the return of similar convulsions in future likely be diminished.[12]

On board ship, Captain Oliver wrote in his journal at the end of the day:

Prosperous, bustling, trading Wellington if not finished is for the present 'done up'. Ships will no more inundate your plains with cattle, and sheep, and horses, from Sidney. Your stock-keepers are no longer alarmed at the inundation of imported stock, altho' the sheep were scabby, the horses bony, and whether the black cattle would outlive the effects of the voyage was a toss up of a bad

Fort Richmond and the Hutt Bridge (1845) by Samuel Brees (ATL A-109-030). A policeman and his companion standing on the bridge felt it move up and down as a northward-moving earth wave, generated by the early morning shock on Thursday 17 October, passed underneath it.

half-penny. Money was so plentiful and Contractors, Butchers and the like so wealthy that thirty or forty deaths in a purchase of lately imported 'acclimated' stock, was mere 'moonshine'. Now how altered! Your Wellington capitalists, men who had the chances are, half a dozen years ago 'had not one sixpence to jingle against another', who the other day were cracking jokes, upon the stock auction, drinking their bottled Bass and smoking their cheroots as they outlaid one another. Alas! How crestfallen are they. One little convulsion of Nature alarmed them. They quailed with the perturbed ground they stood on, another and another came, days went. On each day a fresh shock of Earthquake and with it an increased alarm till down came shop and store, Bonded warehouse and Grog shop, Hotel & Tavern, all in one heap of ruin, then might be seen your sturdy fine fellow of yesterday who came out virtually a duffer and by fortuitous circumstances had been raised to the comfortable commercial man reduced lower than he had formerly been in point of character that is lost all the dignity that becomes a man. One cannot help pitying, but one must despise the man who instead of aiding a Wife and family goes about bereft of his senses and shedding fruitless tears. One poor woman has died this day of fright. There are others of the same sex that set a noble example showing a dignified unflinching, with no womanly fear I was going to write, and I do not know what better word to express it.[2]

The poor woman that 'died of fright' was 43-year-old Janet Nicol, who was found by her husband lying on the floor of their Lambton Quay residence at three in the afternoon, apparently in a fit, but in fact dead. Only five minutes before she had served the coxswain of the *Fly's* gig, sent to fetch Oliver and his fellow officers, with a bottle of grog. 'When I went into the room,' said Mr Nicol at the inquest, 'she was lying on the floor amongst broken dishes and water, which must have been capsized by the severe shock of an earthquake; the large cask in which we kept our water having been upset.' The following day at an inquest at Barrett's Hotel, the jury returned a verdict, 'died of apoplexy'.[13]

Writing his dispatch about the week's events to the Auckland-based Governor-in-Chief, George Grey, at eleven o'clock on Thursday night, John Eyre recorded that he was experiencing 'two incessant and alarming tremblings of the earth; what may be terminated, God alone can tell, but every one seems to have a presentiment that it will end in some still more fearful catastrophe than any which has yet taken place.'[14] The shocks continued 'in quick succession all night'.[5]

CHAPTER 4

THE UNCERTAINTY CONTINUES

Friday 20 October dawned fine and was observed with the utmost solemnity. The congregations at the services were unusually large, the devotions were earnest, and the addresses were varied, appropriate, and duly impressive. To prevent alarm at the earthquakes that were still continuing, most of the services were held in the open air.

The preachers were in fine form. The Evangelists, in particular, were keen to use the catastrophe to bring 'scoffers and mockers' back into the fold of the church, and noted that amid the general destruction and loss of life, it had been 'truly delightful to witness the solomnizing and quickening effects produced upon the community by this striking visitation; the careless have been awakened, the slothful have been aroused, and the zealous have been stimulated to increase activity; prayer, earnest and devout, has been all but universal.' They were sure that there had indeed 'been a great awakening – a shaking of hearts as well as of houses. God has been working, and many we hope will be turned permanently to the Lord.'[1] They were hopeful of the future and encouraged Wellingtonians to put the events in perspective and to count their blessings!

> Earth is not Heaven, New Zealand is not Paradise. Every place has its drawbacks; some of one kind, some of another, but we have certainly not more here than falls to the average share of other places that have perhaps fewer advantages. We have a soil containing the richest elements of fertility; a climate salubrious to a proverb; water in abundance, the best and purest on

earth; wood in plenty and variety, droughts are unknown; snow is seen only on tops of the highest mountains; frost is rarely seen, and then only slightly felt; we are equally removed from the extremes of heat and cold; thunder and lightning are rare; it is not oftener than once or twice in the year that we have a thunder storm, and then the clouds are so high and distant that when the peals roll over our heads, there is more of the sublime and less of the terrible than in almost any region of the globe. The weather is, upon the whole, vastly better for most occupations than in Britain; a fortnight or three weeks will often cover all the time lost by bad weather in out-door occupations from a whole twelvemonth.

We have often high winds, occasionally heavy rains, once or twice a year a slight earthquake, but so slight as often not to be observed; we have had an alarming and destructive visitation, and we may have a return of similar kind, but this is an extraordinary and not a common occurrence. All the danger and loss of life and property have been caused by the falling buildings. In all countries men learn to construct their dwellings on the principles best adapted to resist the most hostile elements around them. The three principal forces to be resisted here in house building are high winds, battering rains, and earthquakes; the last is the force most difficult to be estimated; but from all that can be gathered from experience, observation, and tradition respecting the causes now in operation, there is nothing serious to be dreaded in the future; and if buildings are kept low, sufficiently braced and bound together, and kept always in a proper state of repair, we need dread very little real danger from earthquakes, but in this way safely commit ourselves to the care of a wise, watchful, and merciful providence. Let no one, however, suppose, because all danger from earthquakes seem to be over, that they may therefore return to wickedness or continue in sin; for no law in the universe is more fixed and certain in its operation, that that sooner or later sin is always followed by suffering and misery, while holiness always leads to happiness and joy.[1]

In the afternoon, Barrack Sergeant Lovell died of his injuries while his two children were being buried. The earth continued to be shaken by 13 successive shocks described as 'more or less severe, and with slight intervening shocks at intervals'.

The Scotch church, Wellington, (on a windy day) by Samuel Brees. Sited on the 'Beach' (Lambton Quay), it survived the earthquakes and was used as a place of refuge for the homeless (ATL 81801 1/2).

On Saturday 21st, a 'remarkably' fine day with slight shocks still continuing, Wellington's merchants and other business people, concerned for their livelihoods after a significant downturn in business caused by the earthquakes, gathered at the store of Robert Waitt (businessman and amateur naturalist) to decide what measures were necessary for public safety in the continuing crisis and to ask the Lieutenant-Governor to enforce measures that would prevent people leaving Wellington without giving sufficient notice of their intention. It was also decided to ask him to appoint a Board of Survey to implement suitable precautions with regard to damaged buildings that were still standing but were in a dangerous state, 'so that in future buildings may be erected after some well matured and systematic plan'. It was further recommended that the Lieutenant-Governor be asked to ensure that any disposable troops be employed in removing dangerous buildings, and that they be paid at the usual rate by the owners of those buildings. After the meeting, two of the leading merchants, Messrs Hickson and Rhodes, whose Bonded Stores had been destroyed, took the resolutions of the meeting to his Excellency, who quickly gave his approval.

That afternoon, Sergeant Lovell was buried with military honours. The coffin, preceded by a firing party with arms reversed, was carried by soldiers and followed by a company of the 65th Regiment, the regimental band playing the Dead March from *Saul*. The Lieutenant-Governor, Lieutenant Colonel McCleverty, the Brigade Major and several officers of the 65th Regiment joined the procession, as did ministers of the Evangelical Alliance, and a considerable number of settlers. Lovell had seen 38 years of service in the British Army. His death left a widow and two children in Wellington, and a grown-up son and daughter in Sydney. The Reverend Ironside commented that 'he was a good Christian brother, and a most acceptable local preacher, often filling the Manners-street pulpit to the profit and edification of the people. He was fully prepared for the summons which came so unexpectedly. He had preached at Manners-street the previous Sunday a most excellent sermon from St. John xvii, 4; "I have glorified thee on the earth; I have finished the work which thou gavest us to do", a most fitting theme for what proved to be his last sermon.'[2]

Because of the large number of people who had taken refuge on board the *Subraon*, Reverend Ironside was asked to perform the Sunday service there. The weather was fine and clear with wind from the north, and shocks occurred every three or four hours throughout the day. Passions were running high and religious fervour strengthening at the unexplained continuation of the earthquakes – so much so, that after the shipboard service a Mr Hinchcliffe approached 'a lady of the Hebrew persuasion, who had sheltered there from the terrors of the dread visitation', and told her that unless she renounced 'her faith, and acknowledged the divinity of Jesus, there was no hope of salvation for her'. Arousing a fear that some might seek to use the Jewish community as a scapegoat for the 'evils' that had befallen Wellington, such untoward behaviour brought a quick response from 'an Israelite' that was published in the *Wellington Independent*:

I do not know, Sir, if this individual is, or is not, ordained; but this I do know, that he has exhibited himself to be an insolent and ignorant pretender to the faith he professes . . . Without entering into any religious disquisition, and with the most profound respect for the sincere professors of all revealed religion, be their creed what it may; I have often heard with intense regret, that some of the Dissenting Ministers from the Church of England . . . level shafts of imprecation and revilement against those of the Ancient Covenant I follow . . .

The letter ended with a plea for religious tolerance during such a stressful time.

At the beginning of the week, the safety measures that had been agreed on Saturday were quickly implemented. Colonel McCleverty issued orders for the removal of dangerous buildings. A notice was posted at the Custom House in the morning, stating that a list of the names of the passengers and persons on board any of the vessels anchored in Lambton Harbour would be required from the captain of each vessel 48 hours previous to a clearance being granted to leave Wellington. Eyre considered that 'a regulation of this nature became absolutely necessary in the circumstances of the Colony to prevent persons taking advantage of the occurrence of the earthquake to abscond without paying or making provision of their debts'. Enforcement of the measure was entrusted to Captain Oliver. The regulation was not, however, enacted without objection. Pens came out in defence of citizens' rights, and one – 'Pro Bono Publico' – fired off a letter to the *Wellington Independent*:

Sir, – When the Inhabitants of Wellington presented a Memorial to his Excellency the Governor-in-Chief, on the 11th September last, claiming Representative Institutions, the following paragraph appeared exceedingly objectionable to many of the settlers:-

We beg to remind your Excellency that we at present exist under a form of Government more absolute than that of any other dependency of the British Crown with the exception of Norfolk Island [a penal colony], etc.' Is that paragraph correct or not? To answer this query, allow me to remind you that during the panic created by the earthquakes, several of our liberal influential merchants waited upon Lieut. Governor Eyre, and prevailed upon that officer, 'by and with the consent of the Executive Council,' to promulgate a decree, forbidding all vessels to leave prior to the Captains, or Agents, furnishing the authorities at the Custom House with written notice of departure, 48 hours before the day appointed for sailing, which said notice the authorities placard the door of the Custom House with for public inspection. And furthermore, parties desirous of quitting Port Nicholson are compelled to notify their intention in a similar public manner. Now, Mr. Editor, let me ask is this the boasted privilege of British subjects? And does not this act prove that we are ruled as despotically as any one colony of the crown 'with the exception of Norfolk Island.' In proclaiming such a course, his Excellency may have been actuated with a desire to prevent debtors escaping from their claims, but cannot creditors demand protection from the law? And knowing full well that the law must do so, I hold that enforcement of such a proclamation is not only tyrannical but unjust. Why acting on the same principle what guarantee is there that our liberal 'influentials' may not prevail upon his Excellency to issue a further notice, forbidding under pain and penalties any one in the place, from quitting the settlement for an indefinite period. Such proceedings justifies the assertion, that liberality is all well enough with a certain class in name, but that it does not enter into their composition when they imagine that there might be the remotest chance of loss coupled with the term.

Whether the inhabitants will remain quiescent under such a thorough convict discipline is yet to be seen.

During Monday, only a few shocks were felt and hopes were entertained that 'the subsidence of the Earthquake might be considered as certain'. Monday night passed quietly with only a few slight quivers and things continued to look hopeful as a fine morning dawned. Business began getting back to normal. The military were busy taking down the shattered dwellings and 'activity prevailed in every direction'. Surely, people thought, the shocks must be at an end. But alas, not yet. At ten minutes to two there was a slight tremble, and at five minutes past two a severe shock 'passed through the earth', followed within a minute or two by a second and a third, the last with a vibration that continued for seven and a half minutes. On into that evening the shaking continued with little respite, destroying the confidence which the fine weather and the weakening of the earthquake shocks had begun to restore.

At the time of the first shock, Captain Collinson was standing on the site of the barracks near the wrecked Colonial Gaol, debating with Douglas McKain whether his plans for a new brick barracks should be safe enough to implement, when 'a roll of thunder was heard underground and at the same instant the hill shook as if a gigantic railway train was rushing under our feet and the gaol shook like an aspen leaf'.[3,4] The earthquake instantly decided the question of the barracks' construction in favour of wood.

Throughout the night and into Wednesday, slight shocks continued at an average interval of ten minutes. They were usually preceded by a hollow rumbling noise similar to the concussion produced by the firing of heavy guns. In many instances this noise was heard without being followed by a shock, while the earth continued to tremble the whole time. The meteorological table kept by Captain Oliver on board HMS *Fly* records the incessant shocks felt from Wednesday 25 October to Friday 3 November:[5]

Wednesday 25th . . .	A.M. shocks at 12.5, 12.40, 12.50, 1.10, 2.20, 2.30, 2.55, 3.50, a severe one at 5.20, slight ones at 7.9. P.M. slight shocks at 8.15, 9.5, 11.5.
Thursday 26th . . .	A.M. a slight shock 12.55, rather severe shocks at 1.30, 3.35.
Sunday 29th . . .	A.M. slight shocks at 1.35, 2.30, 3.0, 3.10.
Tuesday 31st . . .	P.M. 7.30 severe shock of an earthquake.
Friday 3rd . . .	A.M. shocks of earthquake 3.5, 3.10, 3.50

A dispatch of Lieutenant-Governor Eyre on 20 November continues the record of principal shocks between the 8th and 19th of November:[6]

8 – Nov.	1 p.m. Heavy shock with grating sound underneath
15 – "	1 p.m. Another smart shock
16 – "	¼ past 3 p.m. Heavy shock
" – "	Midnight – heavy shock
19 – "	¼ to 10 p.m. Slight shock

and he optimistically noted that 'many sharp shocks have been from time to time experienced but have been too insignificant to be worth recording; neither have any of the above been of sufficient violence to do any damage.' It appeared that the violent earthquakes were at last beginning to peter out. In an upbeat mood, the Lieutenant-Governor was happy to report that

> the inhabitants seem now to have quite recovered from the alarm which the continuance and anxiety of the shocks had at first occasioned – Chimnies were in many instances rebuilt and others only repaired from the difficulty of getting bricklayers – new wooden buildings are rapidly taking the place of the mud and brick ones which were destroyed, and in fact a general activity prevails in the town which is strongly indicative of the restoration of public confidence.

CHAPTER 5

NORTH AND EAST

At the time of the earthquakes, Captain Thomas Bernard Collinson[1] of the Royal Engineers had lately returned from a trip up the Wanganui River. He had found the scenery delightful and found the river

> very winding and through a mountainous country. It is about 80 yards broad and on each bank the hills rise 300 feet high and covered with brushwood; at each bend of the river there is a low point, generally with a pa on it, surrounded with a beautiful laurel shrub; the opposite bank being a steep cliff.

For three days he and his companions had worked their way up the river through such scenery.

> Every two or three miles we met a small rapid, some of which we had to pole up and some we charged with paddles, the boys singing and keeping in time to their native songs. Then we came to the stronghold of the river, where several of the pas are collected together containing several hundred people. Here they killed two pigs and roasted them whole for us, and as we visited one chief after another, it was one continual feast for our boys. I think they mean to be friendly and that if nothing new occurs to excite their suspicions, that there will be no more trouble in Wanganui.

But trouble of a sudden and unexpected nature was soon to come. Back at Wanganui, Collinson was suddenly woken early on Monday morning

> by feeling the house tremble with a rumbling sound as of distant thunder and then it shook backwards and forwards as if some giant had laid hold of a corner of it to shake it to pieces. After a minute or two it died away quivering. The

sentries were thrown down and the blockhouses moved as if they were going to walk off altogether.

The movement appeared to come from the northwest like a wave rolling on its course, but leaving a tremulous lateral motion that continued for a long time, resembling the quivering of a dish of jelly. The shock had lasted nearly four minutes and it was almost an hour before the movement ceased. The sensation made everyone feel sick and many suffered frequent headaches. They were due in part to a very disagreeable sulphurous smell that pervaded the air, at times nearly suffocating them, and causing a thick black scum to form on any steel that was exposed.

Remarkably, despite the violent shaking, little damage was done in Wanganui apart from the spilling of water and milk vessels and a medicine bottle or two. A few chimneys had cracked but all remained standing despite being rocked to and fro; the baker's oven was slightly injured. Walls survived intact, and plaster was uncracked except for some that had come down, along with a few bricks, in the church at Putiki on the other side of the Wanganui River. Large quantities of bitumen were washed up on the coast, some pieces of a considerable weight. As elsewhere, the shocks continued for several days and Wanganui residents also felt the larger ones on Tuesday afternoon and early Thursday morning. It was generally agreed that the earthquakes that had rocked the town five years before in 1843 had been more severe. Then, nearly all the chimneys had been thrown down; the brick gable end of the church fell; the earth opened in 'fearful' rents, emitting noxious gases which caused much sickness; the land on the south side of the Wanganui River was elevated; and the channel of the river in places was considerably deepened.[2]

Nevertheless the latest earthquakes had also caused some changes. On the south side of 'the Bluff' near the mouth of the Wanganui River, a rock had been thrust up in the deepest part of the channel to stand 18 inches above the low-water mark; a shoal formed where there had been a depth of six feet, and a long spit of land was elevated along the river bank where previously there had been deep water. The gunboat wharf had noticeably sunk, and a row of posts near the wharf had been pushed outwards a full three feet. Several fissures in the river were also noticed after the shocks, but these rapidly filled up. Beacons at the entrance to the river were much out of perpendicular and a large tree, which had been embedded in the clay and was used as a landing place in the township, had been shifted several yards downstream from its former position.[3]

Going up the Wanganui River on 25 October, the Reverend Richard Taylor saw a great number of landslips, although none of any great size until he came to Hikurangi Pa, where he learnt that the early morning shock on Thursday had caused a large landslide. It had narrowly missed the pa but had covered a number of the cultivations nearby. At Operiki Pa, he found that the missionary teacher, Haimona, had had a very narrow escape during the tremor on Thursday morning. He was sleeping at the root of a large rata tree and woke up when he heard it break as it was shaken rapidly to and fro. He had only just managed to get away, snatching up a few of his clothes, before the tree fell.[3]

The settlement of Wanganui on the Wanganui River in 1848 by Reverend Richard Taylor (ATL E-296-q-159). Despite the violent shaking there was little damage to buildings during the first great shock of the earthquakes early in Monday morning, although the sensation made many people feel sick. With respect to the military 'blockhouses' on the two hills (Rutland stockade on the right; York stockade on the left), Captain Collinson wrote: 'The sentries were thrown down and the blockhouses moved as if they were going to walk off together.'[1]

The Colonial Government barracks at Paremata Point at the entrance to Porirua Harbour north of Wellington, January 1848, by Godfrey Charles Mundy (ATL A-161-016). Completed in August 1847, the five-storey stone building, especially the southeast tower, was severely cracked during the 1848 earthquakes. It had to be abandoned, being locked up and used as a powder magazine, with the troops housed in wooden huts nearby.

Coast near the Ohau River viewed towards the south showing Kapiti Island in the distance by Samuel Brees in 1845 (ATL E-070-012). Reverend Richard Taylor was told that during the earthquakes more than a dozen circular craters each about a yard wide and more than 14 feet deep formed on the beach – the products of liquefaction during shaking. One man fell into one 'and had some difficulty getting out again'.[3]

Further north along the coast at Waimate, the Reverend Woon reported that the shocks of the Monday morning earthquake were 'fearfully' felt. 'His house rocked and reeled under them like a ship at sea, in a gale of wind, and the bells of the chapels rang as though struck for service. His wonder [was] that the frail houses built of reeds and rushes, did not come down upon them'.[4] Further north, at Warea, situated on the coastal lowland west of Mount Taranaki, the German missionary William Riemenschneider heard that at 4 am on 18 October the tide rose higher than it had ever risen before, flooding a migrant camp and washing away a woman (who was later rescued), together with her hut.[5] This was the same day that the sea had risen above its normal high-tide level and flooded part of Lambton Quay and Te Aro in Wellington. Whether this unusual tidal rise was the result of a tsunami generated by one of the aftershocks, or simply a storm surge caused by inclement weather conditions, is unknown. In Wellington, Judge Chapman thought 'the continuance of S.E. gales would, in the absence of earthquakes, be enough to account for this, yet it seems to have produced a good deal of alarm',[6] probably because it occurred amidst the continuing earthquakes. Writing from New Plymouth, the Taranaki correspondent for the *Wellington Independent* (8 November) seems to bear Chapman out, stating that 'on Wednesday [17 October] we had one of the heaviest gales of wind ever experienced here from the N.W., accompanied by much rain and a remarkably high tide'. In New Plymouth, Dr Peter Wilson at the Colonial Hospital found the earthquake

> of more than ordinary vibration . . . Yet it was not in sufficient force to cause the slightest accident or injury to any building, though all our chimneys are of stone, earth, or brick, and there exists more than one clay-built house, so ruinously tottering indeed as at least to disgrace the appearance of the town.[7]

On the 18th, the mail arrived in Wanganui bearing news of the destruction in Wellington. Taylor though it right to appoint a day of public thanksgiving, 'and it is truly remarkable I selected . . . the same day [Friday 20th] which was appointed to be kept as a fast in Wellington. I am happy to say we had a very good congregation.'[3] The accounts of the number of damaged houses so alarmed Captain Collinson that he started off immediately on horseback for Wellington. The further south he went the more apparent the earthquake damage became. At Waikanae there were numerous cracks in the earth. On reaching Porirua, he found the five-storey stone barracks built by the Colonial Government at Paremata Point, with walls two feet thick, cracked from top to bottom.

Collinson was told that the first shock on Monday had commenced with a subterranean rumbling sound. It was followed immediately by an oscillation which shook the building sideways in an east–west direction and lasted altogether about a minute. The rocking was so strong as to throw water out of the roof gutters. Almost

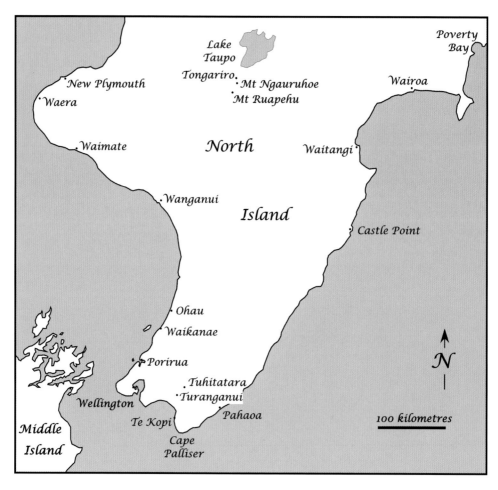

Map of southern part of the North Island of New Zealand showing locations mentioned in text.

all the junctions of the stone and brickwork had been cracked. The second shock on Tuesday afternoon, although not so violent, opened and extended the cracks, and cracked the main wall on the east side. The third shock on Thursday morning, being more violent than the second, widened the cracks still further, and cracked the partition walls in the upper storey at their junction with the main walls and on each side of the large doors, causing the soldiers and officers to evacuate the building. As Collinson sat on a sandhill discussing all this with several of his fellow officers, the same rolling sound he had heard in Wanganui swept over them and the hill quivered for a few seconds.[1]

Reverend Richard Taylor (ATL 1/2-C-14302). Based in Wanganui, Taylor was a prolific recorder of many aspects of New Zealand life, Maori culture and natural history, much of which was published in *Te Ika Maui (New Zealand and its Inhabitants)* in 1855. He recorded the effects of the 1848 earthquakes in and around Wanganui and thought that 'the state of the atmospheric pressure on the earth has something to do with them'.[13]

In early November, Richard Taylor also travelled to Wellington, noting the effects of the earthquakes. Passing through Ohau on the coast, he met a Frenchman who told him that 'during the late earthquakes more than a dozen circular holes made their appearance on the beach in that vicinity; each was about a yard wide and more than 14 feet deep. They were filled with sand and water and raised above the level of the surface.' A man had fallen into one up to his middle and had some difficulty in getting out again. Further south at Waikanae, Taylor lunched with Major Durie, the Resident Magistrate at the police encampment, who afterwards took him to see some of the cracks formed by the

William Colenso, missionary, naturalist, ethnologist and explorer (ATL 3630). Although he does not appear to have left a record of the effect of the 1848 earthquakes at his Waitangi Mission station in Hawkes Bay, while travelling to Wellington in November Colenso found 'much commotion among the natives . . . respecting the Late earthquakes' and saw their effects on the coastal hills.[10]

earthquakes that Collinson had seen two weeks before. '[T]hey run N. and S. and are about six inches wide; there were also circular holes in this place from which large quantities of gas escaped. About 2 miles from Waikanae a large fissure of near 8 inches width opened on the side of a hill.'[3]

Over on the east coast of the North Island, at Wairoa, the Reverend James Hamlin described Monday morning's earthquake as 'severe' and lasting longer than any he had previously felt. He recorded 'another earthquake' on Thursday morning,[8] although the *Spectator* of 28 October was content to report that in 'Ahuriri [Napier], Poverty Bay and to the north, the earthquakes do not seem to have been felt at all'.

William Colenso, at his mission station at Waitangi, Hawkes Bay, at the time of the earthquakes, mentions nothing. Walking down the coast to Wellington in late November, he called at the newly built hut of clay and boulders belonging to Thomas Guthrie and his wife at Castlepoint and was told that the chimney had collapsed during the earthquake.[9] Further south at Pahaoa he found 'much commotion among the natives here respecting the Late Earthquakes and the very heavy flood of the winter; and, above all, some exaggerated reports from Wellington concerning the loss of human life during the earthquakes thereafter'. Reaching the southeastern tip of the coast he saw for himself the effects of the earthquakes:

> all around Cape Palliser, and indeed for many miles on the coast, great alterations had been made by the recent shocks of Earthquakes and several floods; in many places, streams of stones had descended from the very summits of the hills to the plains at their bases, which stones being newly broken were very sharp indeed, making it a painful task to travel over them – especially for my baggage bearers.[10]

Arriving at Te Kopi, the little settlement and harbour on the eastern side of Palliser Bay, Colenso heard confirmation of the earlier report of 'great injury having been done to the town of Wellington by the earthquakes', and now believed it.

Turning inland, Colenso travelled up the eastern side of the Wairarapa Valley to visit the Maori settlements at Turanganui. From the Wairarapa, Maori mission teachers Horomona Pa[11] and Rihara Taki[12] had sent *Koreneho* (Colenso) letters in October describing the effects of the earthquakes. Both mentioned that, during the Thursday morning shock, fire and smoke were seen coming from the ground near Joseph Kelly's house just north of the Turanganui River – the result of spontaneous ignition of methane gas liberated from a swampy area during the shaking. They also told Colenso that the pa of the important Wairarapa chief Te Hiko (referred to by Colenso as Tama Hikoia), a little further north at Tuhitarata near the Ruamahanga River, had been destroyed and that 'the people fled outside with their boxes and everything else'. Just before the earthquakes, they had lost all their crops during a flood. The 28 October issue of the *Wellington Independent* reported:

> From Wairarapa we learn that the shocks of earthquake had been felt rather severely in the valley, though no damage had been sustained excepting the upsetting of milk and cream pans. The natives were dreadfully alarmed, forsook their pah, and took to the bush. They all assert that none of them remember such severe shocks.

Copy of a letter written in Maori describing the earthquake from the missionary teacher Horomona Pa at Turanganui (lower Wairarapa Valley) to William Colenso at Waitangi (Hawkes Bay) dated 20 October 1848 (ATL 910169-1/8-08-01/02). The relevant text and translation are:

'. . . ko tenei wenua ko Turanganui. No te ru na ka kitea he kapura e nga Pakeha e puta ake ana te paowa i te wenua heoi ka mataku nga Pakeha ko matou ko nga Maori ki te tahitaha ko matou kahore I kite atu heoi anake ta matou I kita ai ko te pakarutanga I te pa o Tami Hijoia raua ko Raniera te Iho ko te Tatau . . .'

'. . . This area is Turanganui. In regard to the earthquake, a fire was seen by the Pakeha with smoke issuing from the ground. Consequently the Pakeha were frightened. We Maori, off on one side, did not see this. We saw only the destruction of the pa of Tamai Hikoia and Raniera te Iho ko te Tatu . . .'[11]

CHAPTER 6

THE MIDDLE ISLAND

On 4 October, Thomas Arnold, son of Dr Matthew Arnold of Rugby School fame, boarded the little cutter *Petrel*, and set sail from Lambton Harbour for Frederick Weld's sheep station at Flaxbourne on the Kaikoura coast of the Middle Island. Arnold planned to go to Nelson and start a grammar school, at the suggestion of Alfred Domett, the Colonial Secretary. Weld, whom he met in Wellington, had volunteered transport, with an invitation to stay a while at his station before going on to Nelson. Once out into the turbulent waters of Cook Strait, Arnold had a fine view of the snow-covered Kaikoura Mountains. They steered for Cape Campbell and coasted south for eight miles to a small cove that provided a sheltered anchorage and entry to the Flaxbourne River, up which the *Petrel* could navigate for some distance.

Weld's station (prefabricated, transported from the Wairarapa and erected a year before) stood about a mile from the beach – a white wooden building of two wings with a connecting veranda set in front of white limestone hills with a grassy, almost treeless landscape stretching away before. Two rushy lakes near the house were home to teals, and there were plenty of various kinds of ducks in swamps and lagoons near the sea. The run had 12,000 sheep, jointly owned by Weld and his two cousins, Charles Clifford and William Vavasour, who were then in England. Life appeared as comfortable as one could expect and the larder was in constant need of replenishment with mutton chops and damper, which were served up for breakfast, lunch, and dinner. No milk or butter but plenty of tea, claret, and wildfowl, and an abundance of pickles and sauces.

Left: Thomas Arnold, son of Matthew Arnold, Headmaster of Rugby School in England (ATL 1/2-036955-F). Visiting New Zealand with the intention of staying, he had his first earthquake experience with Frederick Weld at Flaxbourne Station, where 'every timber in the house creaked, groaned and trembled'.[1]

Below: The house of Frederick Weld at Flaxbourne on the Kaikoura coast, Marlborough. View looking northeast towards the limestone peak of London Hill (Canterbury Museum, FAW3). Apart from cracking the kitchen chimney 'right through', the first great shock at 1.40 am on Monday did little damage. The shepherds fled the house, but Weld and his English visitor, Tom Arnold, remained in bed despite the 'almost constant quivering and slight shocks at intervals'.[1]

Apart from farm chores, the days passed with reading, exploring the dry, grass-covered hills and the beach, fishing, and bird shooting. On one occasion, while walking along the river to a small lake about two miles away (Lake Elterwater), they came upon the whitened bones of Te Rauparaha's warriors who had been slaughtered by the Ngai Tahu under the leadership of southern chief Tuhawaiki during a inconclusive skirmish in 1833 or 1834. The skulls all showed 'the deep clean cut of the tomahawk'.[1]

On the night of Sunday the 15th a stormy southerly was blowing when Arnold was awakened 'between one and two a.m.' by his bed being shaken. His first impression was that the wind was shaking the house, but Weld cried out from the next room, 'An earthquake!'

> For about a minute the bed shook violently from side to side; every plank in the house creaked and rattled. The bottles and glasses in the loft above kept up a sort of infernal dance, and most of them fell. The dogs barked and the shepherds who slept in one wing of the house, imagining that the end of the world had come, rushed outside. When the shock was past, there came a few spasmodic heavings, like long-drawn breaths, and then all was still.[1]

Arnold and Weld decided to remain in the house but could not sleep because of the almost constant quivering and the slight shocks at intervals. For Arnold, who had never before experienced an earthquake, 'the sensation produced was singular and awful, its chief element being the feeling of utter insecurity, when that which we familiarly think of as the firm and solid earth was thus heaving and rolling beneath us'.[1]

At daybreak, the shepherds ventured back into the house and found that the stone kitchen chimney was cracked right through. Later, walking around outside, Weld and Arnold discovered that the river flat fronting the house was criss-crossed with long deep cracks, and that rocks had been detached from the sea cliff by the shock, and were scattered over the beach. Arnold supposed that because there had not been an earthquake for many months before, the shock they had experienced was more violent than usual. Six days later, on a beautiful Sunday morning (22 October), Arnold and

Weld walked to the top of the narrow conical hill behind the house for the sake of the view. Seated on the top, they were admiring the snowy seaward Kaikouras when another shock occurred and they distinctly saw the hill on which they sat 'heave and rock to and fro'. Curiously, after the earthquakes, grey plovers made their first appearance at Flaxbourne and were jokingly dubbed 'earthquake birds' by Weld.[1]

Arnold had now been at Flaxbourne for three weeks, and mindful of the inconvenience that his stay might be causing Weld, decided it was time to move on to Nelson. So on Wednesday (25 October), they embarked in the *Petrel* and after six hours sailing in good weather arrived at Cutters Bay, Port Underwood, where they were welcomed by Sarah Dougherty. There they learnt of the havoc created by the earthquakes. The Doughertys' house had been severely damaged. Mrs Dougherty described the first great shock arriving with a hollow, roaring noise. As china and glasses crashed and splintered on the trampled clay floor, the whitewashed, calico-lined clay walls of their cob cottage split and a large crack appeared in their great stone chimney. On succeeding nights they had all slept well way from it.[2]

Elsewhere in Cloudy Bay, not a chimney had been left standing and all the clay houses were more or less destroyed. In the lower part of the Wairau Valley, the raupo house of Henry Gouland, a retired British official from India, had been thrown off its piles on the bank of the Opawa River.[3] Parts of the lagoon and swamp area south of the Wairau River mouth had apparently subsided through compaction, which improved the navigability of the river, allowing the schooner *Triumph* to be the first vessel across the treacherous bar at the entrance on 25 November. The Maori

had reported 'an eruption' to have taken place at the Bluff (White Bluffs) halfway between the mouth of the Wairau River and Cape Campbell.[4] Further up the valley, at Waihopai, the Honourable Constantine Dillon fancied that he had been the greatest sufferer in the settlement,

> for a new house and dairy which had altogether cost me about 80 or 90 pounds has been levelled to the ground. However, the dairyman made the best of a bad job and has put up a hut which he has built on the same spot again and is quite happy. You see that this colonial life, if it does nothing else, gives elasticity of view and teaches people to help themselves. [5]

That night a long red streak in the sky made a brief appearance before vanishing altogether. A few small shakes punctuated an otherwise quiet night. With good company and food, Weld and Arnold stayed one more night before pressing on for Queen Charlotte Sound. They sailed across Port Underwood to Oyster Bay, where Weld sent the *Petrel* round into Queen Charlotte Sound while he and Arnold walked over the hills to Moturangi following Missionary Ironside's track. There they rejoined the *Petrel* and sailed to the pa at Waitohi at the head of the Sound (the future Picton) where they were told that the earthquakes had also been severely felt, with 'the clay houses and the brick chimneys being all more or less destroyed'.[4] The anxiety caused was such that that an open boat sailed over to Wellington from Queen Charlotte Sound in rough seas with several European women on board who were greatly 'alarmed' at the continuing convulsions.

The Wairau Plain by William Fox, January 1848, shows the lower Wairau Valley viewed from the south across the Vernon Lagoons (foreground) with the mouth of the Wairau River in the background (ATL C-013-004). During the 1848 earthquakes, the area of the lagoons subsided, improving the navigability of the river mouth.

The Honourable Constantine Augustus Dillon (ATL NON-ATL-0069). He was living in the upper Wairau Valley, Marlborough, at the time of the earthquakes, when his new house and dairy, which he had built for 80 pounds, were 'leveled to the ground'. Writing to his mother in England in January 1849 about the earthquakes, he remained optimistic: 'You see that this colonial life, if it does nothing else, gives elasticity of view and teaches people to help themselves.'[5]

On the morning of Sunday 29th, Weld and Arnold headed back to the Wairau Plain where they wished to visit the scene of the 1843 Wairau Massacre at Tuamarina, and were shown the spot by the Maori. Here, they learnt that the earthquakes had opened a large fissure near their cultivations. That night another heavy aftershock threw Te Rauparaha out of bed, spraining his hip, and at first light he and his followers hurredly left Tuamarina down the Wairau River and crossed Cook Strait to their base at Otaki. They felt that the earthquakes were a judgement on them for the Wairau Massacre[6] five years earlier on 17 June 1843, when 22 Europeans and between 4 and 9 Maori had been killed over a land-surveying dispute. Calling at Porirua en route, Kanae, a chief with Te Rauparaha, told the Maori there that the earth had opened, the hills were thrown down, water was bubbling up through the cracks, and that the Wairau was 'done for'. The 1 November issue of the *Wellington Independent*, which reported all this, 'whilst not placing implicit confidence in Maori "yarns"', nevertheless felt there was every probability that the earthquakes had opened a volcanic vent in the neighbourhood of Cloudy Bay.

In the meantime, Weld and Arnold had been struggling through deep raupo swamps to reach the pa near the mouth of the Wairau River. Here, they were greeted

The Maori chief, Te Rauparaha (ATL A 114-023). During a severe aftershock he was thrown out of bed and sprained his hip while at his settlement at Tuamarina in the lower Wairau Valley, Marlborough. He hurredly departed for his pa in Otaki.[4]

by Captain Daniel Dougherty, who had been at the river mouth supervising the building of a new house when they first arrived at Cutters Bay, and had now come over in his whaleboat from Port Underwood. The next day (31 November) he rowed them further upriver to the 'Big Bush', a solitary wood of dark 'pines' in the middle of the plain where there was a 'wretched whare' for shelter.[1] At this place the effects of the earthquake were very apparent. Along the river bank there were many cracks, some up to two feet wide. They also saw numerous holes up to ten feet deep where sand and water from below had liquefied during the shaking and spurted out, covering everything for some distance around. Weld thought that this was the result of the compaction of sub-surface peaty matter during the earthquake, causing water to spout up from swampy ground 'thru small craters'. About three miles up the Wairau River they observed that there had been unequal sinking of the land, with what had been high banks along the river now being only a little above water level.[7] The following day they finally parted company, Weld and Dougherty returning downriver to the pa and Arnold setting off 'with elastic step' on the track up the Wairau Valley towards Nelson.

The 21 October issue of the *Nelson Examiner and New Zealand Chronicle* reported the effects of the Monday morning earthquake in Nelson:

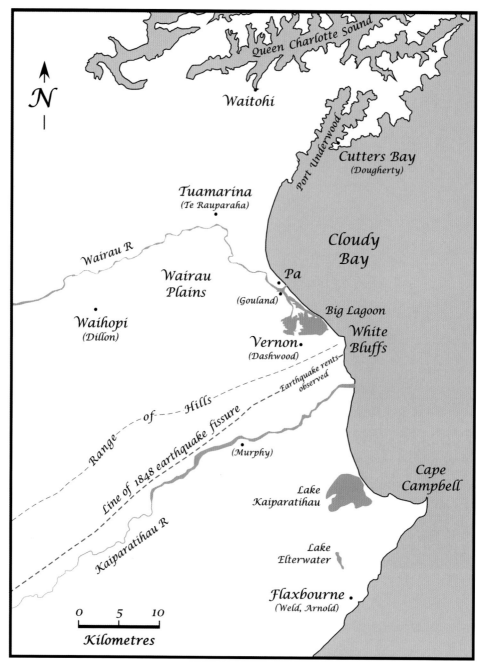

Map showing Marlborough area of the Middle (South) Island. Waitohi is now named Picton, Kaiparatihau River is now the Awatere River, and Lake Kaiparatihau is called Lake Grassmere. The line of the 1848 earthquake fissure discussed in Chapter 10 is also shown, and the area labeled 'Earthquake rents' is from the British Hydrographic Survey map 'Cook Strait' surveyed in 1849 by HMS *Acheron* and published in 1878.

On Monday last, at 1h. 40m., p.m., an earth-tremor, or, as it is commonly called, an earthquake, unceremoniously roused every body from their sleep by the violence and long continuance of its vibratory movements, causing somewhat alarming oscillations of the walls of their dwellings, as well as of the furniture in the interior.

This was certainly the experience of John Saxton and family, who were aroused by a grinding, crashing sound which made them believe that a broken block of wall by their drawing-room corner was falling.

I got up hastily to see when I heard the same grinding and crackling noise extend over the whole house. I wondered if the falling wall was dragging down the whole house. The grinding increased; I felt the floor yield, the bed and wardrobe seemed to be yielding when I thought there was not an instant to lose, and calling Priscilla with baby, John Waring, we got into the dining room and calling up Conrad and Tizzy from the other bedroom, I got a candle from the box. I lighted it at the kitchen fire after nearly five minutes of rocking vibration. The others said it was an earthquake which with my wrong first impression I did not perceive until I crept bare-footed till I saw the ship's lamp still swinging and the clock stopped at a ¼ to 2. This was certainly the longest and strongest earthquake we have experienced. On examination I was surprised and thankful to see that the whole damage was the fall of some loose blocks of plaster in the hall and drawing room.[8]

After being woken by the earthquake at Riwaka, Mr and Mrs Drummond rushed upstairs and took their children outside into an adjoining paddock where they spent the remainder of the night in their nightdresses. Daylight revealed that everything had been shaken from the shelves and the floor littered with broken crockery and glass. Mr Jacks, the schoolmaster who lived next door, had arrived from England only a few months before and had never felt an earthquake. The experience left him so nervous that he did not take his clothes off for three weeks.[9]

The shaking had gone on for about two minutes and some were of the opinion that it was actually composed of three distinct shocks, in quick succession. Over the next hour, there were numerous other shakes of greater or lesser force, and during the whole

John Waring Saxton experienced the 1848 earthquakes in Nelson. They gave him a 'giddy sensation like the approach of seasickness'[8] (Nelson Provincial Museum, Tyree Studio collecion, T1/2 67910/3).

Nelson in 1848 by Francis Dillon Bell (ATL F43142 1/2). The *Nelson Examiner* of 21 October reported that, 'On Monday last, at 1h. 20m., p.m., an earth-tremor, or, as it is commonly called, an earthquake, unceremoniously roused every body from their sleep by the violence and long continuance of its vibratory movements, causing somewhat alarming oscillations of the walls of their dwellings, as well as of the furniture in the interior.'

of Monday, minor vibrations were almost constant. On Tuesday the state of things was no better. Everyone felt the 'violent but short concussion' at 3.40 in the afternoon that did so much damage to Wellington. That evening, John Saxton recorded in his diary that his cottage had been trembling constantly since Monday morning, 'unnerving me and giving a giddy sensation like the approach of seasickness.'[8]

In the evening between nine and ten o'clock, the Aurora Australis was visible for a considerable time, throwing out rays of lurid light, and giving the sky a wild and somewhat ominous aspect – the same display that Wellington's inhabitants thought might have been the effect of a volcanic eruption. The captain of the *Sarah Anne*, which was slowly making her way north from Otakou (Dunedin), had felt the first great shock on Monday morning about 20 miles off Cape Campbell. By Wednesday evening the ship had reached Kapiti Island, where he saw the heavenly display as bright flames shooting up in the sky in a northeasterly direction. At Nelson, the atmosphere was said to have been 'supercharged with electricity'.

On Wednesday the earth still continued in a 'very tremulous state', although no shocks of much force were felt. Samuel Stephens noted that many people 'are much affected by the tremulous movements of the earth, especially those of sensitive constitutions like my own'. The shaking caused 'extreme dizziness, with pulsations of the heart much increased, and great throbbing at the temples'. In some instances it brought on sickness and diarrhoea.[10]

On Thursday, 'a smart shock' at 4.55 am was widely felt in the lower North Island, lasting nearly half a minute, and, as on the previous days, it was followed by a constant succession of vibrations and concussions of lesser force. The Nelsonians were becoming more and more frightened 'and cannot tell at all what to make of it. There are some natives in the town, but they do not seem much alarmed, although they say that they have not in their time known an earthquake last so long.' Vibrations continued all day. At around seven o'clock that evening a brief but violent thunderstorm swept over part of the town from the southeast. Samuel Stephens' cottage was only about fifty yards from the line of the storm and he could clearly hear 'the rain and hail falling in torrents'.[10] By nine the storm had cleared and the Aurora Australis again made its appearance, adding to the 'intensity of excitement and anxiety in the minds of many which had been produced by these unusual demonstrations of terrestrial agitation'.

The combination of the earthquakes and the appearance of the Aurora Australis 'increased the former fright to actual terror among many of the more timid and ignorant of the inhabitants'. The noise caused by the hail during the brief storm induced many to think that a flood was speeding towards the town down the narrow gorge of the Maitai River, causing some of the 'female part of the population' to be 'in a frantic state of fright and terror, screaming and tearing through the streets with their children clinging to them'. On Friday, the earth remained in an unsettled state, with undulatory movements (although generally slight, and more perceptible to people sitting or lying down) constantly occurring at short intervals during the whole of the day and following night. It was noted that all the shocks seemed to have come from the northeast, as they were felt both in Cook Strait and at the mouth of Tasman Bay by the vessels *Supply* and *Ralph Bernal*, and indeed from 'the direction of Tongariro, the principal crater in the northern island'.

The events of the next four days are recorded in Samuel Stephens' diary:[10]

> 21st. Little has occurred to-day different from yesterday, but we have been exempt from any serious shocks, at the same time that the vibrations have been as constant as before. There is no improvement in the people's feelings as to alarm.

> 22nd. Being Sunday, I understand the different places of worship have had a very large accession to the usual numbers of their congregations. The inhabitants seem so fully persuaded that something terrible is going to happen, that they are full of anxiety and restlessness, and consequently, I suppose, anxious to make atonement for their past neglect. At 3.55 p.m. a severe shock reminded us that the phenomena were not yet over.

> 23rd and 24th. There is nothing new to chronicle up to the present moment, but the vibrations are still repeatedly felt at short intervals. During the very early parts of this morning, in bed, I felt several small but distinct shocks, causing the timbers of the house to creak a little. Hitherto all the earthquakes that we have experienced are of that description which I think the Spaniards call 'trembura or tremblura,' and which they do not consider particularly dangerous.

The Nelson newspaper reported that although Nelsonians were considerably alarmed and anxious about their houses and contents, they had not heard of much damage being done except for one or two chimneys partially dislodged, and fissures in the walls of two or three brick dwellings. Although 'undoubtedly the most serious visitation of the kind . . . experienced since the formation of the settlement, and one that will naturally cause a little anxiety to the more timid,' they saw 'no further cause for permanent alarm than may be adduced from the demonstration of other phenomena during nature's progressive development of her mysteries, or, than what may happen in the common chapter of accidents incidental to human life.' The writer concluded with a word of advice:

> it behoves all persons about to erect houses of brick or stone, to be careful in having their work well bonded and cemented together, and the foundations properly secured. Some of the buildings of this kind that have been constructed in the settlement have been far from sufficiently substantial in this respect.

Compared with Wellington, the damage done to Nelson was reported to have been minor. The only building that had sustained any 'serious mischief' was the house of Mr Thompson Esq., which was built on the beach overlooking the Waimea River. The house had been previously damaged by a landslide that had moved it bodily a short distance towards the beach. The earthquakes had thrown down the chimneys and cracked the walls, which were built of large stones collected from the beach. Other casualties with 'slight' damage were the Wesleyan Chapel, the Bonded Store of Messrs Morrison and Slanders, the upper storey of the flour mill, two houses in Trafalgar Street and a mud house near the Eel Pond, one chimney down and several damaged; in all a damage estimate of only £30!

> Indeed, it was wonderful, considering the violence of the shocks at Wellington, and their apparent severity here, that more mischief was not done to us, when we know how slight some of our buildings are, and that many of the chimneys in the settlement are built with clay instead of lime. A gale of wind, such as sometimes experienced on the coast of England, would have produced results far more serious. The late phenomenon, like other secret workings of Nature, is beyond

our comprehension, but as both from European and native testimony it appears to be unprecedented in the memory of man, we see no reason for future anxiety. As an evidence of the electric state of the atmosphere since the commencement of the shocks, it has been remarked in many instances that vegetation has made a progress truly wonderful.

This report in the *Nelson Examiner* of 28 October, however, was at odds with information conveyed to Wellington on the *Mary Ann*, which arrived on Monday 22nd. The passengers stated that they had felt the shocks 'very severely' in Nelson and that most of the brick buildings and chimneys 'were level with the ground'. Alfred Domett, the Colonial Secretary, was obviously aware of this when he wrote to Major Matthew Richmond, Superintendent of the Southern Division of the Province, on 25 October concerning permission to erect a gaol in Nelson:

Sir,

With reference to the authority conveyed to your honour for the erection of a Gaol at Nelson, I am directed by His Excellency the Lieutenant-Governor to request that, should the Earthquake have been sufficiently strong in your settlement to crack or injure brickwork, you will not on any account cause the Building to be constructed of Brick. In Wellington it is not intended to put up any more buildings of that material.

Your most obedient servant,
Alfred Domett, Colonial Secretary. [11]

The earthquakes continued to be felt in Nelson but with lesser and lesser frequency. By 10 November, Samuel Stephens writes 'I do not feel that the effects of the Phenomenon are quite over – because we now and then feel a certain agitation of the earth; which no one can mistake such has been felt in the lower part of the town this very day to a considerable extent, as I am told by credible persons.'[10]

Reports of the earthquake damage inflamed the rivalry between Nelson and Wellington. After receiving the Nelson papers up to 11 November, the *Wellington Independent* of 23 November congratulated 'the colonists of Nelson on their trifling loss', but took issue with their rival's conclusion that 'Wellington was the centre of the shocks, and that they radiated from thence until their force was expended'. It was maintained that the shocks appeared to 'come from the South West, and at Wairau, Queen Charlotte Sound and other places on the Middle Island', where 'they were felt more severely' than at Wellington, and concern was expressed that their 'friends' at Nelson had been 'singularly active in spreading reports as to the extent of damage sustained at Wellington'.

One hundred and seventeen passengers arrived at Nelson on the *Bernicia* from London on Sunday 17 November, heralding 'a revival of emigration'. It was noted that they were told, presumably to put many of them off the idea of settling in Wellington, that all the houses in Wellington 'were shook to pieces', and that the settlers 'were flying away in terror' from the town. This unfounded rumour drew a strong rebuke from the *Wellington Independent* of 25 November:

> Wellington has certainly suffered from this visitation, and though £30 will not repair our damages, the passengers by the *Bernicia* can testify to the falsity of the statements made to them at Nelson. We would suggest to our friends the propriety in future of confining themselves to the truth, for when we hear that such remarks have been made of Wellington, we may reasonably question the truth of the statement that £30 will cover the losses sustained in Nelson. Certainly the passengers in the *Bernicia* estimate the Nelson settlers but lightly for the very questionable accounts received there of the effects of the late calamity here.

On the east coast of the Middle Island, the 'French, German and English' settlement at Akaroa, on Banks Peninsula, had also felt the first earthquake severely, and the shaking was estimated to have lasted three to four minutes; although curiously at nearby Port Cooper (Lyttelton), the proposed port for the New Canterbury settlement planned on the northern side of the peninsula, it was scarcely felt at all.[12] The earthquakes were not felt as far south as Otago, where the local Maori said that such things were unknown, and one man who could speak Maori was unable to make them understand what he meant when he attempted to explain what an earthquake was.[13]

CHAPTER 7

RAP ON THE KNUCKLES

On 19 October, John Eyre, Lieutenant-Governor of the Province of New Munster, sent his first dispatch to Governor George Grey in Auckland. He was evidently in a state of high anxiety after the third violent shock early that morning, and the report was replete with emotion and uncertainty, reflecting the immediacy of the continuing disaster. Eyre was of the opinion that

> a blow has been struck at the prosperity, almost at the very existence of the colony, from which it will not readily recover.
>
> Terror and dismay reign everywhere . . . the energies of all seem paralysed . . . no one has been able to feel for a moment that even life itself is secure and during that period . . . the terror which so frightful a visitation naturally produces in most men's minds will, I apprehend, drive from the colony all who can find the means of getting away.[1]

Grey received the dispatch, which was sent by Maori runner, three weeks later and was dismayed and angry at what he took to be such a defeatist attitude (besides which, he did not like Eyre in the first place). He is reported to have said, 'If Eyre has temporarily lost his nerve, the only thing is for me to go down to Wellington and take over,' and to have started preparations to go in the Government steamer, *Dido*.[2]

But Eyre was no defeatist. Exploring the Australian continent between 1840 and 1845, he overcame sickness, starvation and exhaustion with stubbornness and bravery, reinforced by a strong belief in his own judgement. He had a poor relationship with Grey, who found him annoyingly argumentative, and who constantly reprimanded him and curbed his authority. Initially, this was to the chagrin of the Wellington settlers, who thought Eyre should have had more authority than he had been allowed by Grey, but eventually they concluded from Grey's actions that Eyre was a liability, a pretentious, quarrelsome man, who quibbled over money matters and was of little worth. Yet Eyre had conducted himself well during the earthquakes, showing none of the ineptitude and indecision unfairly attributed to him by Grey.

In his first report to Grey regarding the earthquake, Eyre was reacting to the moment, with little time to contemplate, and his views were an accurate reflection of those of the majority of Wellington's inhabitants on Thursday the 19th. When Eyre wrote his second dispatch on the 21st, reporting the continuation of the shocks, further damage, the solemn fast-day, and news from elsewhere, he was no longer inclined to 'hysterical-sounding' passages.[2]

Almost immediately on the receipt of Eyre's reports, Grey sent them, together with his own assessment, to Earl Grey at the Colonial Office in London.

Edward John Eyre, Lieutenant-Governor of the Province of New Munster at the time of the 1848 earthquakes (ATL C-016-003). He was accused by many of losing his nerve and being defeatist after writing an initial report, described as 'graphic, but highly mischievous', on the earthquake damage and its effect on the inhabitants of Wellington.

Government House, Auckland, Nov. 13, 1848

My Lord,

It is with extreme regret that I transmit for your Lordship, information copies of two despatches, which I have received from Lieut. Governor Eyre, detailing the great loss which has fallen upon the inhabitants of Wellington from the effects of a series of severe earthquakes.

Having received verbal information from various sources to a much later date than Mr. Eyre's despatches, I shall perhaps best meet your Lordship's wishes by stating generally the full effects, as far as they are yet known, which these earthquakes have produced upon the whole colony.

They have been severely felt throughout the whole tract of the country, extending in the Northern Island from New Plymouth to Wellington. It is thought, also, that a slight shock was felt in Auckland early one morning. In the Southern Islands earthquakes have been felt from Nelson to Cloudy Bay. In the southern portion of this island the shocks commenced on the 16th of October; and upon the 31st of the same month, to which my intelligence extends, shocks at Wellington were still said to be of frequent recurrence; and early this month a few shocks were still felt at New Plymouth or Taranaki.

The only place at which serious damage appears to have been inflicted by this visitation was at Wellington. There, unfortunately, three lives were lost by the falling of buildings, and I fear that a large amount of property must have been from the same cause destroyed. I am told that the total loss of property at that place is estimated at £50,000; of which loss, however, a great portion would fall upon the Colonial Government, the public buildings having been those which were principally destroyed: wooden houses and buildings, I am informed, escaped without any injury whatever. Still, however, admitting that there may be much exaggeration in the estimated amount of loss, and that a great portion of it will fall upon the Colonial Government, there can be no doubt that the loss of property which has taken place will be severely felt by the inhabitants of so young a settlement, which has already had so many difficulties to contend against. The most serious loss to Wellington, however, is the want of confidence which has been created, and the panic which has arisen, under the influence of which I understand that many families intended to abandon the place.

This feeling will, however, I trust, speedily subside; and in order that every encouragement and assistance may be afforded to the sufferers, which it is in the power of the Government to bestow, I propose to proceed to Wellington with the least practicable delay, where your Lordship may rely upon my doing all I think you would desire to be done for the relief of the inhabitants of that place,

The intelligence I have received of the effects of this earthquake at Taranaki and Wanganui, quite satisfy me that up to the latest dates Wellington was the only settlement at which any injury of consequence occurred, or was likely to take place.

I have &c.,

G. Grey [1]

Governor-in-Chief Sir George Grey in 1850 (ATL G-623). He was annoyed by the seemingly defeatist attitude of Lieutenant-Governor John Eyre's first dispatch describing the effects of the earthquakes and decided to go to Wellington and take matters in hand. He arrived aboard the HMS *Havannah* on 20 November, putting Eyre in a humiliating position.

Published in the Auckland newspapers, Eyre's dispatches provoked a sharp rebuke from the editor of the *New Zealand Editor and Southern Cross*:

> Of the expediency, if not absolute necessity, of his Excellency's presence [George Grey] in the South, there are few, if any, conflicting opinions. 'The restoration of public confidence and *the whole future prosperity* of Wellington' do most unquestionably depend upon the promptitude, the energy, and the address exhibited in remedy of those disasters which have overwhelmed the town, and which – judging by his dispatches, prophetic of further evil, and cowed by those which have already occurred – have filled the mind of the Lieutenant-Governor [Eyre] with maudlin terror, and guided his pen with drivelling prognostications of desolation and gloom. Instead of animating the despondent, he has been the first to give added bitterness and unseemly exhibition of his own. If the Man quailed beneath the unwonted terrors of his position, the Governor was imperatively called upon to rally his energies, and to cheer and inspire the people confided to his rule, rather than prostrate and damage their settlement by expressions, ex cathedra, such as these, wafted on the four winds of heaven, and, which if calmly weighed over, are, after all, but the sorry evulgations of a nervous individual. 'A blow has been struck (quoth the despatch writer) at the prosperity, almost at the very existence of the settlement, *from which it will not readily recover*'.
>
> Did Lisbon recover from her blow [the great Lisbon earthquake of 1755]– compared with which, Wellington was but the filip of a ladies glove?
>
> 'Every one seems to feel a *presentiment* that it will end in some still more fearful catastrophe than any which has yet taken place. The terror which so frightful a visitation naturally produces in most men's minds, will *I apprehend* drive from the colony all who can find the means of getting away.' We hope like Burchill in the *Vicar of Wakefield*, we may, without profanity, exclaim – '*Fudge!*'
>
> When Terentius had withdrawn his shattered legions from one of the most sanguinary fields of the ancient world, such was his well grounded confidence in the energy and constancy of Rome, that in reply to *his* despatch, detailing his defeat, he received the thanks of the Senate, who pronounced him to have *deserved well of his country* because he had not *despaired* of the affairs of the republic. We would council Lieutenant-Governor Eyre to study Terentian

maxim, recommending him to pause ere he again indulges in doubtful predications and injurious predictions.

There is *stuff* enough in Port Nicholson to defy despair; and energy sufficient to repair the calamity under which they labour. May their arms be strong – their hearts resolved.[3]

Within a month a further rebuke was published in the *Nelson Examiner and New Zealand Chronicle*:

One of the most extraordinary documents we have ever seen in print, is Lieutenant-Governor Eyre's despatch to the Governor, furnishing an account of the earthquakes at Wellington. We have reprinted it, that our readers may have the benefit of perusal. If the object of his Excellency had been to destroy what the earthquakes had spared, he could not have taken means more likely to ensure success. His despatch is graphic, but highly mischievous. When men are called to high stations, they are expected to be equal to all emergencies, and, if they do not rise with the occasion, they show themselves unequal to their task. It was quite proper that the Lieutenant-Governor should communicate the fact of the earthquakes at Wellington, and the extent of the mischief they had caused, to the Governor-in-Chief, but knowing that his despatch would become a public document, and that it would carry with it an unusual degree of authority, he should have been careful how he made statements that would be productive of mischief. Only fancy the effect this document will produce in England, where it was forwarded by the *Dido* almost immediately after its receipt in Auckland:– 'Wellington in ruins,' 'an immense destruction of property,' 'melancholy loss of life,' 'numbers of persons ruined,' 'many persons afraid of remaining in their houses at night, and retiring for safety to the bush among the hills in wild and inclement weather,' 'a blow struck at the prosperity, almost at the very existence of the settlement from which it will not readily recover,' 'terror and dismay' everywhere prevailing, 'energies of all the settlers paralyzed,' 'a general presentiment that some still more fearful catastrophe will happen,' and his Excellency's apprehension that the calamity 'will drive from the colony all who can find the means of getting away.' Here is comfort for the friends of the colony in England. Here are inducements offered by a Governor to people to come and live in his province. And what an encouraging effect it will have on the mercantile interests of Wellington. However readily people in England will consign goods, and have business transactions with the ruined inhabitants of a town destroyed by an earthquake, where all who had the means will have left the settlement. The actual losses incurred by the Wellington settlers have already been sufficiently great, and needed not this most ill-advised dispatch of Lieutenant-Governor Eyre to crown their sufferings.[4]

The Wellington papers remained quiet on the matter, perhaps because they had been in the thick of it and understood what it was really like. There was probably also an inkling of sympathy for their chastised Lieutenant-Governor, 'a good sort of a person, only rather wanting in tact, and very anxious to do the right thing by everyone'.[5]

By this time Eyre had penned his third report on the 29th. Apart from conveying the news of the wreck of the *Subraon*, this dispatch was more in line with Grey's expectations, depicting a Wellington that was by this time regaining some semblance of normality:

Awful as the visitation was during its continuance, and calamitous as have been the results, there are yet many circumstances of consolation and encouragement in connection with it. First, such convulsions appear to be most rare. No similar ones have taken place since the colony was established, nor can I ascertain that the natives or others ever remember any such violence and long continued duration. Secondly the worst shocks have not been the first, and thus a timely warning has been given to quit brick or other dangerous buildings, and little loss of life has ensued. Thirdly, not a single wooden building has been destroyed or, as far as I am aware, even injured, and thus amidst all the alarm and apprehension which so sudden and fearful an occurrence naturally excites, places of shelter and security have existed for the whole population, and no other real injury has been sustained by a large number of the inhabitants than has been occasioned by the breakage of fragile articles in their houses. Fourthly, there is no doubt whatever that not a single brick building in the town has been really well and properly built, so that it is impossible to say how far brick buildings, if really well and substantially put up, would have withstood the violence of the shocks experienced; even as it was some one or two buildings of brick have been left, comparatively speaking, uninjured.'[1]

The report reached Auckland on 18 November, the day that Grey was officially invested with a knighthood, and he quickly sent it on to Earl Grey with another covering letter.

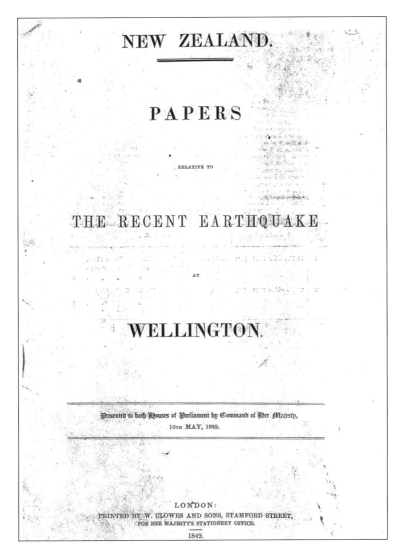

NEW ZEALAND.

PAPERS

RELATIVE TO

THE RECENT EARTHQUAKE

AT

WELLINGTON.

Presented to both Houses of Parliament by Command of Her Majesty,
10TH MAY, 1849.

LONDON:
PRINTED BY W. CLOWES AND SONS, STAMFORD STREET,
FOR HER MAJESTY'S STATIONERY OFFICE.
1849.

Title page of official reports concerning the 1848 earthquake from Lieutenant-Governor Eyre to Sir George Grey, who sent them to Earl Grey at the British Colonial Office. The reports were presented to the British Parliament on 10 May 1849.

Government House, Auckland, Nov. 20, 1848.

My Lord,

Your Lordship will be glad to find from this despatch, that although slight shocks of earthquakes continued to be felt at Wellington upon the 30th of October, yet still, that public confidence was gradually returning, and that there was every reason to suppose that the greatest violence of the earthquake had subsided; the information I have received from various parts of the island, lead me to believe that no serious damage has occurred at any other place than Wellington. I intend to proceed to that settlement to-morrow, and I trust that the measures which have been already taken, and which I propose to take upon my arrival there, will soon entirely restore public confidence, and that its former state of prosperity may, under the blessing of Divine providence, be speedily re-established.

I have &c.,

G. Grey. [1]

Grey's second dispatch to England arrived a day later than the first (25 April 1849) and both were published and presented to both houses of the British Parliament on 10 May.

Eyre's last dispatch, written on 10 November, indicated that things had at last returned to normal, but by this time the Governor-in-Chief had started for Wellington on the 18-gun warship HMS *Havannah* under the command of Captain Erskine. He was accompanied by Chief Epuni from Petone, who was with Grey at his investiture. Grey's arrival in Wellington on 20 November put Eyre in a humiliating position, and from then on things between them grew even worse.

CHAPTER 8

GENEROSITY DECLINED WITH THANKS

On the evening of Thursday 9 November, Auckland business people and church leaders gathered in the Hall of the Mechanic's Institute for the purpose of expressing sympathy with and providing aid to the suffering Wellington community. Theophilus Heale, Esq., surveyor and Judge of the Native Land Court, occupied the chair. His lordship, the Anglican Bishop George Selwyn, was present, together with the Reverend Henry Lawry and other Wesleyan ministers, and 'all were earnest in their expressions of good will'. It was unanimously agreed:

> That this meeting has heard with feelings of the most painful kind, of the calamitous results of the earthquake which has been experienced in the districts adjacent to Cook's Straits, more particularly in the settlement of Wellington, where several lives have been lost, and a large amount of property has been destroyed; and as our fellow Colonists in all parts of New Zealand, are regarded as brothers, bound together, in many respects by a unity of interest, though occupying separate localities, rendering the welfare and prosperity of each community matters of great concern; this meeting does most deeply sympathize with the inhabitants of Wellington, and Cook's Straits, under their present great affliction.[1]

Before the meeting had ended, upwards of £266 had been subscribed, with further collections to be made following services the next day at the Wesleyan Chapel.

Two weeks later, on the evening of Thursday 23 November, another public meeting took place in the Britannia Saloon, the main entertainment theatre in

The Anglican Bishop, George Selwyn, in 1849, by George Richmond (ATL C-016-019). Selwyn headed the list of Auckland donors of money to relieve the distressed in Wellington following the 1848 earthquakes, with a donation of £100.

Wellington, convened for the purpose of receiving the address of sympathy from the inhabitants of Auckland, and to consider what steps should be taken with the money subscribed there for the relief of cases of extreme distress, if such existed, occasioned by the late visitation.[2]

The meeting had been called for half-past seven, by which time the venue was filled to overflowing. Described as being 'the most numerous and respectable assemblage yet held in Port Nicholson', all sects and all classes were there; the Episcopalian, Roman Catholic, Presbyterian, Wesleyan, Independent, and Primitive Methodist ministers occupied the platform together with numerous officials, merchants and storekeepers.

On the motion of William Hickson, Justice of the Peace and merchant, and with loud applause and cheers, the Anglican Minister, Reverend Robert Cole, was asked to chair the meeting. Expressing his 'great pleasure in being present . . . for the object of the meeting was to follow out the glorious principles of Christianity

Reverend Robert Cole (ATL F110428 1/2). On the Public Fast Day (Friday 19 October), he was busy preaching sermons, two in the morning and two in the afternoon, after which he went and conversed with the Maoris at Te Aro Pa 'in lieu of sermon'. On 25 November, he chaired the public meeting to decide whether or not to retain the money raised in Auckland for 'the relief of cases of extreme distress, if they should exist, occasioned by the late visitation'.[2]

– good will to all men', he then read the account of the public meeting in Auckland, and told the gathering that the Committee appointed by the subscribers to manage the appropriation of the fund had received two amounts: £400 from the inhabitants of Auckland, and £100 from the Bishop, the money being conveyed to Wellington by Sir George Grey aboard the HMS *Havannah*. After expressing his admiration for the generous sympathy of the northern settlers, Reverend Cole called upon Rabbi Abraham Hort J.P. to move the first resolution of the meeting.

'The inhabitants of Wellington have great cause for thankfulness,' said Rabbi Hort, for 'in the midst of all our afflictions, at the time of the greatest danger, we must be thankful to the Supreme Being that so few lives have been sacrificed.' Also affirming his 'great pleasure' to be at the meeting, where everyone could put aside their political sentiments, and express themselves 'in the warmest possible manner, to the settlers at Auckland, for their very generous and truly noble conduct', he proceeded to read the following resolution:

> That the Address of the inhabitants of Auckland to their fellow-colonists in this
> settlement, has excited in our minds, feelings of the highest esteem, and the
> most unfeigned gratitude.

'I fully concur in all that has fallen from the very worthy chairman and the last speaker on the conduct of the inhabitants of Auckland, and I wish to second the resolution,' said Reverend Samuel Ironside, and being put to the chair it was carried unanimously.

Father Jeremiah O'Reilly had been deputised to propose the next resolution. He read it to the meeting, after insisting at some length that he could 'scarcely find language' to do so, and ascribing the generosity of Aucklanders to the 'dispensation of Providence' rather than merely human good nature.

> We, the Inhabitants of Wellington and its vicinity, in public meeting
> assembled, return you our most grateful and cordial thanks for the sympathy
> and benevolence you have displayed towards us under the recent severe
> dispensation of Providence with which this settlement has been visited. While
> we acknowledge that this calamity has been, in some instances, attended with
> great destruction of property, we have abundant reason to be thankful 'to the
> great Being by whose inscrutable Providence this visitation has been permitted
> and controlled,' for his many mercies, particularly that so few lives have been
> lost, and that the injury sustained by the settlement, is ascertained to be so
> much less than was anticipated. It is also a great consolation to us. In this trial,
> to receive from our fellow-colonists at Auckland, such ready sympathy, so noble
> and magnificent a proof of their regard; and we trust that the spirit of Christian
> charity, of which you have given so memorable an example, may ever influence
> the colonists of New Zealand, and join in the very bond of peace the different
> settlements into one prosperous and united colony.

Adoption of the address was duly seconded by the Presbyterian minister, Reverend John Inglis, who, like the others, could not 'speak more highly of the deep sympathy so feelingly displayed, and of the liberal assistance so promptly rendered by our fellow colonists in Auckland'. However, he reminded the gathering that the news received in Auckland 'had evidently produced a false impression' because, on the contrary, the situation in Wellington was 'highly favourable, the public health has never been better, sickness of any kind being at present unknown, – employment is plentiful and wages good – provisions are abundant and cheap, and every thing connected with the season and the settlement encouraging'. Nevertheless, commending the kindness of the Aucklanders, he had no hesitation in seconding the address 'and it was carried forthwith without a dissenting voice'.

As the applause died down, Mr Jonas Woodward, pastor of the Congregational Church, rose to announce that he had been instructed by the Committee appointed to dispense the relief money to tell the meeting what it had done. He reiterated that 'the Committee regarded themselves as trustees of those Auckland friends who had made donations and were anxious to hear the opinion of the meeting'. He said that the Committee were prepared not to release the money 'if it should appear that there were no cases of distress as would justify its application'. So far hardly any such cases had been reported to the Committee and they were of the opinion that the money should be returned. That many persons had been much inconvenienced, that many houses contained more than their ordinary tenants, they were quite aware; but while they deeply sympathised with distress in every form, they still thought all such cases might be fairly met by efforts on the spot. Nevertheless, he made it clear that the Committee had not yet made a decision, preferring to leave this to the present meeting. The important matter of what steps should be taken with the money 'subscribed for the relief of cases of extreme distress occasioned by the earthquakes' was now opened to the meeting for discussion.

After a short interval of muted discussion, Mr Joseph Roots stood and said that he objected to returning the money, and thought that it should be held for a short period. The chairman scanned the hall for a seconder but no support for the motion was forthcoming. In the lull that followed, the Wellington notable and New Zealand Company supporter Dr Isaac Featherston rose to speak. 'I had not intended to take part in the proceedings,' he began,

> but I cannot permit such a resolution to be mooted without saying a few words
> in reply. I acknowledge in the strongest manner the obligation under which the
> inhabitants of Wellington have laboured to the Auckland residents, for the people

at the North had come forward and exhibited the kindliest feelings towards this place. But it appeared to me that the money has been subscribed under a false impression. The Auckland settlers were told in a certain despatch, the sender of which you all know [Lieutenant-Governor Eyre], that 'Wellington was in ruins' and that the inhabitants were labouring under severe distress. Our fellow-colonists have contributed a considerable – nay a very considerable amount to alleviate the destitution which they were led to believe existed in this place. That despatch further went on to state that 'a blow has been struck at the very existence of the settlement from which it will not readily recover.' Why gentlemen, since the formation of the colony, has not this settlement struggled through and overcome more serious calamities than the late affliction – serious though that affliction has been? And was it proper for a man holding so high a position as the writer of the despatch in question, to speculate on possible contingencies, and to predict the ruin of the settlement? That despatch has been published at Auckland, and it created an impression in the minds of the Northern settlers that the inhabitants of Wellington were suffering extreme distress and misery.

Under these circumstances he thought it would be unjust for Wellingtonians to retain the subscribed 'monies' and proposed that, despite 'their heart-felt appreciation of the prompt and munificent manner in which aid had been afforded', it would be better to refuse to take 'improper advantage of the kindness' and to advise the Committee to return the money subscribed 'coupled with the expressions of their warmest thanks'.

Robert Stokes, editor of the *New Zealand Spectator and Cook's Strait Guardian*, seconded the resolution, but Robert Carpenter, a member of the Legislative Council, interjected, warning of the 'impropriety of doing anything which, by any probability, could be construed into insult', and proposed an amendment supporting Mr Roots' earlier suggestion 'That this meeting instruct the Committee to accept the funds so kindly subscribed by our fellow-colonists at Auckland'.

Promptly seconded by Mr Roots, the amendment and resolution were then put from the chair and this time it was carried almost unanimously. Featherston, who was one of the few that voted against the resolution, now stated that he wished to modify his earlier proposal by suggesting a further resolution:

Dr Isaac Earl Featherston (ATL 1/2-005051). He was a strong supporter of the New Zealand Company. At the public meeting to discuss what to do with the support money sent from the citizens of Auckland, Featherston thought that it would be unjust for Wellingtonians to retain the subscribed 'monies' and that it be returned 'with the expressions of their warmest thanks'.[2]

In allowing that a few cases of distress might exist, I believe that if any one in the community has suffered to any extent, that such distress would already have been made known. However, there was already a fund of about £80 – a balance of a former subscription not appropriated, which might be employed in alleviating the distress of anyone who can establish a case, and, though I have no idea that distress, really speaking, exists, I would still like to propose:

That the same Committee be requested to receive applications from parties who have suffered by the earthquake; and if they find that the case required it, that they set on foot a public subscription in Wellington for the relief of such sufferers, and that the Committee have power to add to their number.

William Fox, principal Agent of the New Zealand Company, now spoke to the resolution. While he thought that the conduct of the Auckland settlers had been 'noble and manly' and 'manifested the warmest feelings of humanity', he felt that if the money was retained, 'a great set of injustice would be committed'. He regretted the fact that the people of Auckland had been led to believe that conditions in Wellington were far worse than they really were, and as to the belief that the settlement would not recover from the calamity, he 'would ask, has it not partially recovered already? And as to its ultimate recovery' he and many others had 'not a shadow of a doubt'. 'If distress exists,' he insisted defiantly, 'we will alleviate it from our own means,' and in doing so strongly supported Dr Featherston's motion.

The Reverend Ironside, however, was not convinced that this was the best thing to do and preferred a middle course. 'Although I do not desire to appropriate the fund,' he said, 'I would rather retain it for a short period, and should distress be found to exist, alleviate it out of the money so generously placed at our disposal.'

Having listened to all the arguments for keeping or returning the money, Mr Woodward, who had presented the views of the Wellington Committee, now took the opportunity of stating his own. After hearing the opinion of Mr Fox, he felt that the retention of the money could not be justified even on the ground of expediency.

> All of you know that there is something like eighty pounds which could be made available, and this, with such private aid, as I feel confident would be afforded, would meet all the cases in which relief might be desirable. I earnestly protest against using money which has evidently been contributed under an erroneous impression.

'What would be said,' he warned, 'if it was found that we allowed distress and suffering to exist for six weeks, and only thought of relieving it when we had other people's money to do it with!' He reminded the meeting that 'it had been proposed to make application to the British Government, and it evidently would damage any claims which must necessarily be for a large sum, by accepting these donations, which can not reach the heavier sufferers, and are not required for the lesser.' The arguments had persuaded him that the Committee should be advised to return the money forthwith.

The hour was growing late, and when the resolution was put from the chair it was carried unanimously. The money was to be returned with thanks. Wellington would stand on its own. A vote of thanks was passed to the Reverend Cole 'for his upright and courteous conduct in the chair', and the meeting concluded in an orderly fashion.[1]

As a postscript, the decision of the meeting led to one further suggestion by a Wellingtonian that was published in the *Wellington Independent*:

> the gentlemen whose names are found mentioned as having charge of the distribution, might lend the money received in sums varying from 10 to £20 upon the promissory notes of the borrowers, payable on demand, without interest, it being understood that the money would not be required for two years, or such further period, as might be arranged. In this way persons of limited means, but in good employment, might be assisted in the reconstruction or repairing of their houses, and the money when ultimately returned to Auckland, might be used in erecting some public building commemorative of the occasion and of the benevolent spirit which led to the collection of the fund.[2]

CHAPTER 9

A NINE DAYS' WONDER

The sequence of great shocks that occurred early on Monday morning, on Tuesday afternoon and early on Thursday morning of that third week of October 1848 caused the progressive destruction of Wellington. The tremors had been felt from about the latitude of Banks Peninsula in the south to that of New Plymouth in the north, and were felt most strongly in the vicinity of Cook Strait, their violence gradually lessening in a northwesterly and a southeasterly direction from there. They had all been of a similar character: a sound like subterranean thunder, accompanied by a vibration of the ground for a few seconds, and then a rapid heaving oscillation of the earth, which, over an interval of a few seconds, died away with a quivering motion. Judge Chapman noted that 'when the shocks occur during a heavy gale, as on Monday, [their] dull rumbling sound is not perceptible: it is overcome by the greater noise of the wind. When the shocks occur in calm, they are generally preceded and sometimes followed by a strong puff of wind.'[1]

These shocks, and that on the afternoon of the 24th, had been succeeded by a great number of smaller ones, varying from ten to twenty per day. They continued, gradually decreasing in number and force, throughout the remainder of October and well into November. These small shocks had not much of the heaving motion of the larger ones, but felt more like the firing of cannon immediately underneath the ground, and were sometimes so frequent that they sounded like a 'distant cannonade' and the earth appeared to tremble incessantly for two to three hours together. Although no one had counted all the shocks in any one day, Chapman thought that they must have exceeded one thousand; 'At times there has not been one minute between each – at others, three, four and five minutes, and so on, diminishing to one or two in the house. On three of the days we were several hours without one.'[1] The earthquakes were the most violent he had experienced since coming to Wellington in 1843, but according to the *Wellington Independent* (25 October), the Maori living in the Wellington area had experienced a more powerful earthquake about ten years earlier (before Wellington was founded), that 'occurred during the day, and so violent was the commotion, that many were thrown down on the ground'.

In Wellington, the earthquakes' noise, and the motion of the earth, seemed to some people to come from the south, to others from the north. As damage to buildings occurred principally on both their southeastern and northwestern sides, and the shocks were felt at Nelson a little more violently than at Wanganui, hardly at all at Hawke's Bay, and as strongly at Banks Peninsula as at Wanganui, the earthquake shocks were concluded to have run in a northeasterly and southwesterly direction.

Ground damage in Wellington had been minor despite the severity of the earthquake shocks and the observation that during the Tuesday 17 October shock 'the earth in some parts was moved in waves averaging about 12in. in height'.[2]

> The shocks were most violent and destructive on alluvial sand, or gravelly formations, as [at] Te Aro and Thorndon flats; on the clay-stone formation, as on the terraces surrounding the town, they were somewhat less violent; and on the Karori-road, at Wade's town [Wadestown], and Kai Warra [Kaiwharawhara], where the houses are nearer the rock, the shocks were comparatively light. Small clefts were made in the earth in some places, but the water in the wells has not been disturbed . . . From the Heads round to Cape Terawite [Terawhiti] large rocks were rent, and fragments rolled down and were precipitated into the sea . . .[3]

It was a time of anxiety and helplessness for Wellington's inhabitants in the face of an overwhelming and unpredictable force. The only course of action left was to offer up prayers in the desperate hope that these unprecedented divine visitations would soon end. Most took comfort and strength from this. Everyone was well aware that earthquakes, volcanic eruptions, thunderstorms, disease, and all such 'terrific phenomena' were the effects of natural causes and that they could be more

'Gullies formed by Earthquakes, Wellington, New Zealand 1849' by Charles Gold (ATL B-103-006). The earthquake 'gullies' are presumably the bare parts of hill slopes that represent landslides caused by the 1848 earthquakes, possibly in combination with the wet weather at the time. The view is probably along the Ngaio Gorge.

or less explained by the operation of physical laws. Nevertheless, there was mind as well as matter:

> the mind was considered immeasurably superior to matter; there was a system of laws for the regulation of mind, as well as for that of matter; and the governor of the universe was an intelligent Being who constituted the last link of the chains of both physical and moral causation. So it was that God obtained submission to his physical or natural laws by punishing disobedience, and as matter is made subservient to mind, the lesser must yield to the greater.[4]

Or so the people of Wellington were taught to believe. A strong belief in the Bible and an unlimited confidence in the power of prayer, 'the divinely appointed means of preservation', was regarded an important link in the chain of moral causes, and it was believed that as prayer affected God it must also indirectly affect his control of natural elements such as earthquakes.

Armed with this moral strength, Wellingtonians could begin to think of picking up the pieces, of rebuilding, even when there was nothing to indicate that an end to the earthquakes was in sight. Nevertheless, the daily, hourly fear of another large shock, the inclement weather, the bright lights in the sky, the fact that there seemed to be no connection between the shocks and changes in barometer or thermometer, all helped to exacerbate the uncertainty of the situation. Shelter, security, and self preservation were the prime concerns.

Many headed for the surrounding hills, particularly Maori; others sought refuge on board ships anchored in Lambton Harbour. For the majority that remained in town, living under canvas, or in cramped conditions with others in those wooden structures that remained intact, soon led to friction. People looked to anything that might indicate the end of the earthquakes. The dawning of a fine day was enough to raise people's spirits, and rumours that volcanic activity was releasing pent-up gases, the cause of all this disturbance, from deep within the earth, were the best of news.

If there was to be a future to contemplate, an assessment of the damage was needed, and a semblance of order and purpose had to be imposed. A Board of Inquiry was duly constituted by Lieutenant-Governor Eyre on 25 October. Charged with preparing a thorough report on the earthquake damage as quickly as possible were Robert Park, a civil engineer representing the Wellington settlers, the architect Henry St. Hill, and Captain Thomas Collinson of the Royal Engineers, acting on behalf of the Government. With the assistance of Sergeant Mills of the Armed Police, who had made his own list of damaged buildings, they completed their report in less than a month.

The report[5] noted that almost every chimney in the town had been broken off close to the roof; that the brick buildings which had suffered least were those that had bond-timbers in the brickwork, and had been lined with wood, or weather-boarded; that in a great many houses in which the wall plates had not been carried throughout the gable, the gable-ends had been thrown down, but that the gable-ends of hipped roofs had not suffered so much. They also found that in almost all the brickwork, the mortar had been of a 'very bad description', being composed of lime and clay, instead of lime and sand as it should have been. As a result there had been so little bond in the brickwork that many walls were shaken down in single bricks.

The Board of Inquiry therefore advised 'for greater security against fire than a weatherboard house, a strong wooden frame upon a brick foundation, filled in with bricknogging laid in mortar, and covered outside with strong laths and plaster, and inside with boards of plaster.' But they realised 'that there will be almost always a great many houses in the Town built entirely of wood', and accordingly recommended 'that all future wooden houses should be separated from each other as much as possible, both as security against fire, and because the action of a shock is sometimes of an undulating kind, that will take more effect on a continuous line of buildings than on several detached small ones.' It was also ascertained that the amount of damage done to the Town, 'at the utmost, is not more than £15,000 in property of all descriptions, and that includes £3500 of the Colonial Government, and £1000 of H.M. Ordinance'.[5]

Judge Chapman also made a rough estimate of the losses:

Chimneys	1,500
Clay houses	1,000
Small brick Do.	3,000
6 larger houses and Inns	2,400
Merchants stores and Dissenters Chapels	3,500
Breakages	2,000
Government Buildings	3,000
	£12,800

He noted that Board of Inquiry's estimate of 15,000 pounds was for the town only and probably did not include the breakages inside buildings, and with additions 'for Karori 200 pounds – Porirua road 400 pounds – Hutt 600 pounds', and so on, the total could be more like 18,000 pounds.[1]

William Fox, Agent for the New Zealand Company in Wellington, was eager to allay fears that the earthquakes had affected public morale enough to deter intending immigrants from England. Writing to the Company on 31 October, he was of the opinion that

> all fear of the shocks becoming more severe has ceased, and we know the limited extent to which they actually occurred, all alarm has subsided, confidence in the place appears entirely reassured . . . owners of the damaged buildings are beginning to erect new or repair their old ones, and the event is already regarded as little more than a nine days' wonder.[6]

Referring to accounts of average earthquakes in other countries, and Maori traditions of earthquakes, Fox maintained that he would have no hesitation in emigrating to New Zealand:

> There may, no doubt, be worse in store for the colony, but when it is considered what sharp shocks were felt in London in 1750, and in Scotland for many days in 1839 (not to mention many minor indications there of volcanic action) and that no repetition of severe shocks has occurred in those places for such long periods, we may fairly hope that the late event here has also been of accidental origin, and will prove to be of solitary occurrence.

What was more, he stressed, 'the results of the earthquake had been unimpressive; three lives only having been lost by the falling of a wall, and a large vessel sailing at the very moment when the alarm was greatest, with only about forty people willing to take advantage of the opportunity; and most of the passengers likely to remain.' There was little doubt, he continued, that 'the danger of a voyage by sea, is, in fact, greater than any that we have been subjected to, and probably every one who travels 100 miles on a railway, incurs a greater risk than he would do by living a lifetime in this place.'

The problem was the brick chimneys and buildings of Wellington, so many being 'thrown down, or so much injured as to be unsafe', reputedly because of their bad construction; this alone had given rise to an 'exaggerated view of the violence of the commotion'. The quality of the lime used to bond the bricks was seized upon as the

William Fox, the New Zealand Company Agent in Wellington at the time of the earthquakes (ATL 1/2-017921-G). His report to the company downplayed the effects of the earthquakes so that prospective immigrants to New Zealand would not be discouraged. Although appreciating the monetary support offered to Wellington from the citizens of Auckland following the earthquakes, Fox had 'not a shadow of a doubt that if distress exists we [Wellingtonians] will alleviate it from our own means'.[6]

main problem. The best lime came from the Nelson area, and in Wellington it was both expensive and scarce. Builders had got into the habit of using clay, sometimes mixed with a little lime, in the construction and 'the result has been that the bricks in reality only hold together by their own weight, and a very trifling force is sufficient to dislodge them'.

To prove his point, Fox had visited several collapsed brick buildings, the gaol and the Methodist Chapel among them, and collected some of this 'curious building material, picked off the fallen bricks', which he sent back to Head Office in London. He noted that 'in two or three, traces of lime are discoverable by the eye, and possibly in all by chemical analysis, but some of them appear to be entirely devoid of it'. A small number of chimneys built of 'English bricks and cement' had largely withstood the shocks, but because 'so few Englishmen' had any previous acquaintance with earthquakes, the Wellington settlers had not been able to obtain much information as to 'what may be the most desirable substances and forms to resort to in building for the future'. He was hopeful that the Court of Directors might be able to 'procure any information . . . either from men of science at home, or from practical persons in South America, the Mediterranean, or elsewhere (where earthquakes were frequent)', as this would 'probably be of much use in giving confidence to such as may wish to provide against a similar contingency for the future'.

Fox's letter to the New Zealand Company was promptly published in its mouthpiece, the *New Zealand Journal*. The letter drew a detailed response under the title 'Defective construction of houses in Wellington'[7] published in the same issue of the journal, from an avid reader and strong supporter of the New Zealand Company, Charles Rooking Carter, who was to arrive in Wellington in 1850:

> Sir, – As I notice from the latest despatch of the New Zealand Company's Agent, Mr. Fox, that much of the damage done at Wellington, by the late earthquake, (or rather, earth tremor, for it is unworthy of the name of earthquake,) resulted from badly constructed buildings, perhaps you will permit one daily engaged in the art of building, in all its branches, to offer a few observations on those principles of construction best calculated to resist the effects of that oscillatory, undulating, or wave-like motion of the earth, which occurred at Wellington.

Charles Rooking Carter, builder (ATL 1/2-082367). Although yet to immigrate to New Zealand, he felt duty-bound to publish an article in the *New Zealand Journal* (London) entitled 'Defective construction of houses in Wellington' after reading a report of the earthquake damage written by William Fox, the New Zealand Company Agent in Wellington.[7]

He had 'after a careful examination of facts' no hesitation in pronouncing that the greatest part of the damage sustained by the buildings in Wellington was due to the inferior bricks used in their construction – 'mud mixed with an exceedingly small portion of lime in the proportion of a bushel of lime to two barrowsfull of a substitute for sand' – noting that chimneys constructed of English bricks and cement had withstood the shock. 'Good mortar binds and holds all the different portions of a house together, and it is this binding together that enables a dwelling to withstand an ordinary shock of an earthquake; for one part cannot fall without the other,' he wrote, and then proceeded to explain the best way of constructing with brick.

I would suggest to our friends in Wellington, to lay the foundations of their houses two or three feet below the natural surface of the ground on a concrete footing, and if it is to be a building wholly constructed of bricks, let the mortar be good, which will be the case if mixed in the proportion of one bushel of unslaked lime to one bushel of sharp sand; or even one bushel of lime to two of sand will not make bad mortar; the lime should be stone lime, shell lime and chalk lime are of an inferior quality. The stone lime that slakes quickest is the best; if it be scarce in the neighbourhood of Wellington, it is the duty of the authorities to make a search for some. Sand should be washed clear of earthy or clayey particles. I should think it may be found in abundance in the beds of the rivers. Experience has proved, in England, that sea sand, after it has been dried in the open air during the summer is equal, if not superior, to land sand; and when mixed with lime and coal ashes makes most excellent mortar. The bricks should be of good marl and dried in the sun, and then well burnt.

When the brick work is three feet in height, from the foundation, and is fourteen or eighteen inches thick, lay three pieces of hoop iron on its horizontal top, and at half the length of the wall, and each corner; take out of the surface of the wall, a brick, or two ; and wrap each piece of hoop iron once round one brick, then replace the bricks level with the top of the wall, and carry up the wall three or four feet more; and then again adopt the same plan, and again, till the wall is the height required, and then the hooping should be worked in diagonally; and if the first floor of joists are well spiked to the bond timber, and the ceiling joists to the wall plate, a degree of stability will be insured capable of withstanding the shock of an ordinary earthquake.

Carter was careful to add that his 'hastily penned remarks' were not intended for the instruction of Wellington builders, as 'they, I should say, know their business as well as many in London', but rather were meant to call the attention of the public and local government to the necessity of making some local building Act, as a guide and legal standard for workmen, builders, and architects, i.e., to enact New Zealand's first building code. At the time he wrote this letter, master builder Carter had not lived through an earthquake. He immigrated to New Zealand in 1850 and was living in Wellington when the great earthquake of 23 January 1855 struck. On that occasion mind quickly merged with matter as he rushed out of a hotel where he was sitting and sprinted along Lambton Quay in panic to his house and family. Naturally, he reported that his house and chimneys all survived the severity of the shaking, and was quick to point out that this vindicated the earthquake-proof building measures he had proposed to Wellington's inhabitants six years earlier.

The earthquakes kept on occurring all through 1849, and although usually much less violent, they were at times frequent, especially in January. All were diligently reported by Judge Chapman in letters to his father at 2 Tillotson Place, Waterloo Bridge, London.[1]

> *January 12, 1849*
> Strange to say the slight shocks still continue at the rate of from one to five in the 24 hours. In all December there were only five days on which we did not notice any shocks and even on those days especially during the night shocks may have occurred which we did not notice. This month they have occurred every day hitherto and U think of rather greater strength than those of December. Still as I have before stated out of perhaps more than 2000 shocks in all only <u>four</u> and perhaps even only <u>three</u> were of a destructive character but are quite enough for destruction as Lisbon [during a great earthquake on 1 November 1755] can testify and I am disposed to think that in these numerous slight shocks there is safety.

January 21

Earthquakes – After the 11th we had four days without any shocks that we were aware of. On the 16th a slight shock was felt. On the 19th four shocks in the day; on the 20th 3 or 4 slight shocks at night. These are however nothing.

February 14

This morning at 20 min. past 2, we had a very sharp shock which lasted about 25 or 30 seconds. I think it was sharper than any we have had since October. During this month we have had continual shocks on 10 days out of 24: but others may have occurred which escaped our notice.

May 4

Yesterday we had a very sharp shock at 1 ¼ p.m. another more moderate at 1 a.m. The first stopped the French clock and set all the bells ringing. It is now almost 7 months since they commenced.

October 3

The Earthquakes are now pretty well over though we felt a slight shock and heard a rumble yesterday. Although we have had no one damaging shock since the grand shocks of October 1848 yet I find our chimneys are shaken loose by the constant disturbance of shocks for so many months. I hope we shall be able to repair them without taking down.

Chapman's final letter to his father mentioning the earthquakes was on 16 December 1849, by which time he thought that the shocks had 'been reduced to the average state – or rather less – since the end of August'.

CHAPTER 10

FIRE, WATER, AIR: A FISSURE IN THE GROUND

In 1848 it was thought that the elements of heat, water and air were the essential ingredients of earthquakes. The experiences of people who climbed active volcanoes, such as those in Italy, had 'proved' the connection:

> ascending Etna during an uneasy state of the mountain, we were repeatedly thrown off our legs by the shocks, which, from their loud reports and from the vibratory motion accompanying them, conveyed the impression that pent up air was expanded, or that water was by heat converted into steam, the expanse of which splintered the rocks, produced a report like a park of artillery, and laid us prostrate on the ground.[1]

There were thus good reasons for speculating that the subterranean causes of the earthquake and the volcano were the same. Assuming the presence of reservoirs of liquid lava in the interior of the Earth, it was not difficult to understand how steam might be generated whenever rain or sea water, percolating through the rocks, came in contact with such lava; and how, when the resultant steam was generated, the overlying crust of the earth could be fractured and dislocated. During such movements, it was believed that fissures could be formed and invaded with gas or fluid, which sometimes failed to reach the surface, and at other times was expelled. And when the strain on the rocks had caused them to split so that the roofs of pre-existing fissures or caverns had collapsed, vibratory jars would be produced and propagated in all directions, like waves of sound through the earth's crust, and at various speeds, according the violence of the original shock, and the density or elasticity of the various kinds of matter through which they passed – faster through solid rock than through unconsolidated sediment.

Along with the theory blaming the collapse of underground caverns, the idea that earthquakes were caused by subterranean discharges of electricity had also been promulgated. It stemmed from a fanciful analogy with the noise and shock that accompanied lightning, from the extreme rapidity with which earthquake motion is dispersed, from the perceived 'electrical' state of the air before and after an earthquake, and from the presence of a sulphurous smell which was thought to resemble that produced by an electric shock. The idea was reinforced by the fact that electrical phenomena were common during volcanic eruptions, where they were thought to be produced by the evolution of large quantities of steam 'and other elastic fluids, the decomposition and subsequent regeneration of water and other processes'.[2]

The Reverend Richard Taylor thought there was a subterranean communication between Mount Erebus, a known active volcano in Antarctica, and Mount Ngauruhoe; and he believed that earthquakes, at least in the vicinity of Wanganui, were caused by vast volumes of gas or superheated steam suddenly being formed in this passage, and rushing north towards the vent of Ngauruhoe. Taylor's conjecture was later 'confirmed' by another Wanganui resident who, walking home from the town one evening, heard an earthquake and stopped to note the result. Soon after the shake had passed, he distinctly observed a very bright light and heard a prolonged explosion from the volcano. This was followed after a five-minute interval by an unusually loud bang.[3]

On the evening of Tuesday 17 October, a light was seen to the northeast of Wellington, and again on several subsequent nights appearances like the reflection of some powerful light were seen among the clouds to the south. It was concluded that a volcano was erupting near the centre of the North Island, and that this was causing all the mischief. It was well known that across the centre of the island a chain of volcanic disturbance was in constant activity. In the words of Judge Chapman:

The volcanic region of pumice hills looking towards Tongariro and Ruapehu by George Angus in 1844 (ATL PUBL-0014-28). It was thought that subterranean activity in the central volcanic region of the North Island of New Zealand was the origin of the 1848 earthquakes. The *New Zealand Evangelist* conveyed the hope that the 'safety valve' of one of the volcanoes 'would be opened by means of which pent up subterranean fires, that by their explosive force are shaking the earth in all directions around us, would be allowed to escape, and the return of similar convulsions in future likely to be diminished'.[6] Unfortunately, travellers to Wellington from Wanganui reported that Tongariro had not 'burst out'.

It commences at Tongariro [present-day Ngauruhoe] – a conical mountain about 10,000 feet high, visible from Wanganui, and from Cook's Straits, which continually emits jets of steam and smoke. In January 1845, Te Heu Heu told me that it was throwing out flame. From Tongariro, the chain extends along a line of lakes, hot springs, fissures and steam-jets of a very remarkable character, to the Bay of Plenty, where White Island is an active volcano the crater being near the water's edge. This last I have seen. The direction from Tongariro to White Island is about N. E. Some of the hot-springs must exist under pressure, for their temperature is 216 degrees at the surface. Some of the mud-jets are at the boiling point. One of the lakes is called RotoMahana (Roto, lake; Mahana, warm). Underground noises are continually heard, new openings occur from time to time, and extensive land slips are not uncommon. In 1846, a mud slip destroyed the pah of Te Heu Heu, at Lake Taupo, and he with 50 of his people perished. Such is the normal state of the volcanic district – a very small increase of volcanic action would account for all that we have experienced during the last week.[4]

Chapman went on to surmise that recent heavy rains would have been sufficient cause for such an increase in volcanic activity:

Any extraordinary mass of water suddenly disengaged from its customary channels and basins, and let loose upon these hot vents and fissures, would produce sudden changes in the relative density, elasticity of the air, and steam in the volcanic caverns, and be followed either by collapse, or by great efforts to escape, or perhaps by both. If it be true that Tongariro has become active, i.e., more than usually active, may we not rely on it as a safety valve?

These conjectures were echoed by the editor of the *Wellington Independent*:

Whichever theory we may take as the correct one in respect to the causes which originate earthquakes, it cannot be denied that the quantity of rain which fell during the winter, pouring into the bowels of the earth, and coming in contact with substances which are well known to generate heat and steam, or gas, has in great measure produced the convulsions which we have now to record. But when we consider that the natives themselves were surprised at the nature of the shock, and that not even the oldest of them remembers anything similar in force or duration, we are surely justified in considering the present as an extraordinary occurrence, an occurrence which in all probability may not occur again. Furthermore, we have sufficient evidence to prove that gas or steam has now forced a vent for itself, and a vent which, according to appearance, is not likely to be closed for a long period to come. Knowing this, knowing that in no single instance has a vertical or upheaving motion been experienced with any fatal effect in New Zealand, and knowing from the formation of the country that the effects (or dying efforts) of that power, which first raised the country from the deep and made it fit as an habitation for man, must be experienced in a manner similar to the subsidence of a storm, we consider that taking all these circumstances into consideration, we have no greater reason to dread abiding in New Zealand than we had on the first day that the settlers landed on these shores.[5]

The *New Zealand Evangelist* was equally hopeful that the earthquakes would soon end, but rather less certain:

if the dormant embers of some smoldering crater had been kindled up and burst forth – if some close by volcano was come again into a state of activity, a safety valve would be opened by means of which the pent up subterranean fires, that by their explosive force are shaking the earth in all directions around us, would be allowed to escape, and the return of similar convulsions in future likely be diminished.[6]

and:

The volcanic agency resembles a small thunder cloud that may be violent in the locality where its force is expended, but is little felt beyond that single spot. From all our experience we must conclude that earthquakes in this country are limited in effect and fluctuating in appearance, and the next one may probably be in some locality where this one has been very little felt, but as the causes in operation are so much beyond the knowledge of man, and though regulated by laws as fixed and uniform as any other that God has imposed upon matter, yet as these are known only by himself, it is right for us to speak of probable events connected with earthquakes with extreme caution.[7]

But travellers such as Captain Collinson, who arrived in Wellington from Wanganui, reported that Tongariro had not 'burst out . . . and persons begin to think, that the convulsion came from the S and E'[8] – where there were no active volcanoes. Wellingtonians concluded that the earthquake must have originated on the southwest coast of the Middle Island, and probably south of Cascade Point, where a number of violent earthquakes had occurred in 1826 and 1827. From there, it seemed reasonable to suppose that the earthquake must have travelled in a northeast direction, through the old volcano of Banks Peninsula, and undoubtedly *towards* the closest known active volcanic islands, Curtis (now Kermadec) Islands, about 400 miles off the east coast of New Zealand. However, when the schooner *Harlequin* arrived in Wellington on 27 October after a ten-day voyage from Otago and Port Cooper (Lyttelton Harbour), those on board had seen no signs of a volcanic eruption. Moreover they were able to confirm that the brilliant night lights were those of the Aurora Australis, or Southern Lights, rather than 'meteoric appearances of some kind, arising from the air being supercharged with electricity that were known to be common during earthquakes'.[9] The idea of volcanic activity in the Middle Island as a cause of the earthquakes seemed to be unfounded.

The other important earthquake agent was thought to be the air. Judge Chapman observed his barometer scrupulously during the earthquakes. He noted every fall and rise, together with earthquake times, and recorded the state of the weather. But there was, as he realised in the end, no correlation between periods of low pressure and the occurrence of earthquakes.

The Reverend Taylor, however, persisted with the air theory. 'I have generally noticed that earthquakes are preceded by gales,' he wrote,

> and it has struck me that the state of the atmospheric pressure on the earth has something to do with them. When we reflect that, in all probability, the solid crust of our globe is not thicker in proportion to its size, than the skin of an inflated bladder, atmospheric pressure must have much to do in predicating the angularity of our surface, and any cause affecting it must correspondingly affect the earth it presses upon.

But he had to admit that 'there are other agents at work by which also these fearful phenomena are occasioned.'[10]

Nevertheless, on the basis of the available evidence of the directions from which the earthquake shocks arrived in Wellington, it appeared that the earthquake source lay somewhere to the south. Two years later Frederick Strange, naturalist with the HMS *Acheron*, wrote that:

> From information I was able to gather from different parties who reside at Bank's
> Peninsula, Nelson, Wellington and Entry Island [Kapiti Island] in Cook's Strait,
> I believe that the seat of the earthquake is to be found in the Middle Island,
> between Cloudy Bay and the mountains called by Captain Cook the Looker-on
> [Seaward Kaikouras].[11]

Strange's estimation of the earthquake source area was correct. In fact, it was in a little-known and sparsely settled part of the South Island that evidence for the cause of the 1848 earthquakes was to be found, although it was to be another 40 years before this was realised.

Under pressure from the New Zealand Company to acquire land suitable for sheep farming and to provide rural sections to fulfil the Company's obligations, Governor Grey negotiated with the Ngati Toa people the purchase of a considerable block of land in the Wairau and Awatere valleys in the northeastern part of the South Island. The block became known as the 'Wairau Purchase'. The agreement was signed on 18 March 1847 and soon survey parties set to work surveying the flat land of both valleys. A year later the sections were allotted to purchasers who had bought them in England seven years before, and who now became their absentee owners.

Land outside the surveyed sections was regarded as 'waste lands', and this included a large part of the lower Awatere Valley, or the Kaiparatihau district, as it was then known. In the middle of 1848, the New Zealand Company, having these waste lands at its disposal, decided to issue depasturage licences. The Nelson gentry, whose lands were rapidly becoming overstocked, saw the chance to acquire grazing rights in the Wairau and Awatere as too good to miss. They began selecting runs in early 1848, well before 'The Terms and Conditions for the Depasturage of stock on the Company's waste lands' were advertised in October, and proceeded to drive their surplus and sometimes starving sheep onto them illegally. These wealthy landowners thus became squatters in the 'Wairau'.

The 11 November issue of the *Nelson Spectator* reported the observations of some of the squatters who had been to their runs to see how they had fared in the earthquakes:

We find by the accounts given us by a party of gentlemen returned from the Wairau, that in the Kaiparatihau district the shocks of the late earthquake were very severe. From the White Bluff, extending in an easterly direction [this should be in a *westerly direction* as the stated direction would be out to sea!], there is for several miles a fissure in the ground, and the high and precipitous banks of some of the branch rivers have been thrown down.[12]

We learn more about this 'fissure' from Major Matthew Richmond, Superintendent of Nelson, who visited his run, Richmond Brook, on the south side of the Awatere River in early November to inspect the 1100 in-lamb ewes he had transferred there from Nelson. On entering the Awatere Valley via the bridle track over the hills from the Wairau Plain, he encountered a 'crack' that 'quite straight crossed the country for miles'. His horse had difficulty getting across it, and Richmond was astonished to see that the fissure had passed through an old whare near the track, 'dividing it in two pieces standing 4 feet apart'.[13] Crossing the Awatere River at Taylor's Ford, he came upon the shaken ruin of the cob cottage where one of the shepherds, Daniel Murphy, and his wife had lived.

Major Matthew Richmond, Superintendent of Nelson (ATL 1/2-005393). After visiting the Wairau and Awatere valleys in November 1848, he wrote to Reverend Richard Taylor, saying 'it is quite awful to see the manner in which the earth is rent in many places'.[23]

Pressing on, he arrived at Flaxbourne, the run adjacent to Richmond Brook, and was welcomed by Frederick Weld. For the next three days, Weld showed him over much of his property and Richmond was appalled to see how badly the river flats had been torn up by earthquake fissures. Returning to the Wairau Valley to call on Edwin Dashwood near Big Lagoon, his route took him past some old Maori potato gardens that were peppered with holes, from which sand and water had been ejected during the shaking.

There were also belated reports of earthquake damage further up the Awatere Valley. During their exploratory journey up the Awatere River in December 1850, where they discovered and named Barefell Pass, Weld and George Lovegrove found a

> very extraordinary earthquake fissure horseshoe shaped some 30 ft wide by 12 ft deep – thus – it appeared as if the neck of land on which it was had – so to say, been shaken till one side bulged out leaving a fissure on the ridge – the bottom was covered with strips of sod which appeared to have sunk into the aperture – apparently not more than two years had elapsed since the chasm was made – the

Frederick Aloysius Weld, sheep farmer, explorer, naturalist and later Premier of New Zealand (ATL MNZ-04339-1/4). During a trip to England in 1856, Weld told the great British geologist, Sir Charles Lyell, that during the 1848 earthquakes, a 'big crack was produced in the high range of mountains, from 1000 to 4000 feet in height, which extends to the S. from White Cliffs [White Bluffs] in the Bay of Clouds [Cloudy Bay],' and that he, and others, had traced it 'over an extent of 60 miles, in a N-S direction, on a line parallel to the axis of the range.'[17]

hill was yellow gravelly clay – there is no possibility of its having been formed by the action of a water.[15]

A year later, Stephen Nicholls, travelling to the upper Awatere Valley to look for a suitable run for his employer William Adams (who became the first Superintendent of the province of Marlborough in 1860), described the area that he had chosen, in the vicinity of Upcot and Gladstone stations, as 'all Ravines, Chasms and precipices, and great Chasms made by the earthquakes both frightful and awful to look at'.[15] In a public lecture on the 'Geology of the Province of Nelson' on 1 October 1859, Ferdinand von Hochstetter reported that 'below the confluence of the Blairich river with the Awatere . . . the side of the mountain has slipped with an earthquake rent',[16] the information being reported to him by his colleague Julius von Haast (appointed as Provincial Geologist of Canterbury in 1861), who had accompanied Hochstetter on his geological explorations.

In the succeeding months and years, the big crack or fissure along the Awatere Valley, which was on average no more than 18 inches wide, was traced by Weld and others 'worthy of trust' over a distance of 60 miles, along the flank of the range of mountains that separates the Awatere and Wairau valleys.[17] It was said to be '5 or 6ft. deep and 30ft. broad' in many places,[19] to resemble 'a canal without water for miles on end',[18] to be 'upwards of eighty miles in length resembling a macadamised road and about the same width',[19] and to follow 'a direct line for 85 miles; in some places as wide as a canal, in other places only a fissure in the earth of various depths'.[20] A line of 'earthquake rents' between White Bluffs and Dashwood Pass, marking the line of the Awatere Valley fissure, was marked on the British Admiralty map of Cook Strait surveyed by the HMS *Acheron* in their 1849 hydrographic map; Frederick Weld provided them with the information when officers and crew of the ship visited him at his Flaxbourne station during their survey of Clifford Bay and Cape Campbell.

A small section of the fissure first appeared as a line on a map in January 1854, when a Mr Thomas Musgrave was employed to survey the land purchased by William Aitkinson in the lower Awatere Valley. It was marked on Musgrave's map as a single line extending for some seven kilometres and labelled 'The fissure of the

October 16 1847 earthquake'.[21] Although the year was wrong, the month and day were not, and Musgrave later corrected his mistake on a copy of the map. The 1848 fissure marked on Musgrave's map thus has the distinction (although this was not realised at the time) of being the first earthquake fault to be mapped in New Zealand and possibly anywhere in the world.

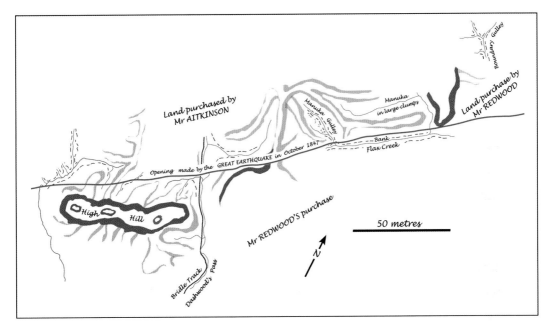

A copy of part of Thomas Musgrave's 1854 survey of the property of Mr Aitkinson, between Stafford Creek and Boundary Stream in the lower Awatere Valley, Marlborough, showing the line of the 1848 earthquake fissure labelled 'Opening made by the GREAT EARTHQUAKE of October 1847 [sic]'.[21] (Scale is added.)

Although geological terminology in the early 19th century equated 'fissures' with 'faults' and with earthquakes, 'fissures', 'cracks' or 'rents' that formed during earthquakes were differentiated from faults (also known as 'shifts' or 'slips') where displacement of rock strata, mineral veins or the land surface could be demonstrated. In his book *Geology and Mineralogy* (published in 1836), the flamboyant Dean William Buckland, Professor of Geology and Mineralogy at Oxford, quoted William Conybeare, another well-known geologist of the time, defining faults as

fissures traversing the strata, extending often for several miles, and penetrating to
a depth, in very few instances ascertained; they are accompanied by a subsidence
of the strata on one side of their line, of (which amounts to the same thing) an
elevation of them on the other; so that it appears, that the same force which has
rent the rocks thus asunder, has caused one side of the fractured mass to rise, and
the other to sink. – The fissures are usually lined with clay. [22]

It was from New Zealand, however, that the British geologist Charles Lyell
made the correlation between earthquake fissures and faults in 1856, based on
the evidence of a '90 mile' long rupture and vertical displacement along what was
to become known as the Wairarapa Fault. Found in the Wairarapa Valley, in the
southern North Island, the rupture had occurred during a great earthquake in 1855.

Lyell also received information about the Awatere Valley fissure formed during
the 1848 earthquakes from Frederick Weld when he visited England in 1856.

Mr. Weld, who was in the Middle Island during the previous earthquake in
1848, tells me that at that time there was produced a big crack in the high range
of mountains, from 1000 to 4000 feet in height, which extends to the S. from the
White Cliffs [White Bluff] in the Bay of Clouds [Cloudy Bay].[17]

Because it could not be established whether the ground surface had been
displaced on one side of the fissure relative to the other, Lyell was careful to refer to
the feature as a fissure rather than a fault.

And there things were to remain until Alexander McKay, the Government
Geologist, mapped a fissure that had formed along the Clarence Fault in the South
Island. He also described a rupture formed by a large earthquake on 1 September
1888 along another South Island fault, the Hope Fault. For the first time, a geologist
was at hand to investigate and record the effects of a large earthquake, and McKay
was the first in the world to discover that the movement along faults during
earthquakes was horizontal as well as vertical. This discovery was to influence his
work tracing the fissure created by the 1848 earthquakes, identifying the Awatere
Fault, and his eventual understanding that long superficial earthquake fissures
were the surface traces of major fault lines.

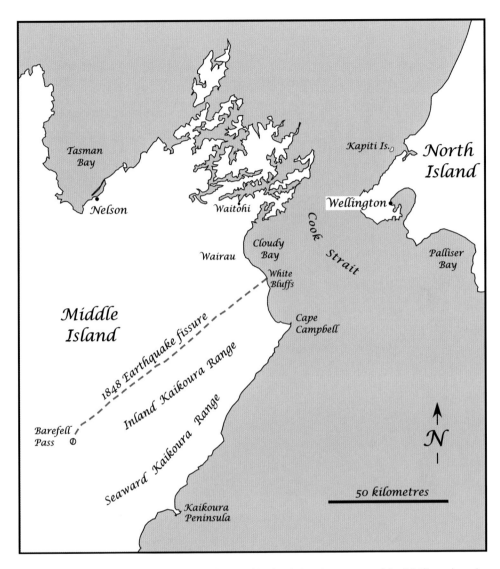

Map of the northeastern part of the Middle (South) Island showing extent of the 1848 earthquake fissure (from White Bluffs to Barefell Pass) along the Awatere Valley, as traced by Frederick Weld and others in the 1850s.

CHAPTER 11

INTERLUDE:
SEARCHING FOR
AN EARTHQUAKE AND
AN IMPORTANT DISCOVERY

The violent earthquake that rocked the central part of New Zealand just after 4 am on 1 September 1888 was the largest since the the 1848 earthquakes and the great earthquake of 23 January 1855. It commenced with a rumbling noise and a few seconds of slight shaking, followed by the main shock lasting 40 to 60 seconds, and was felt from Invercargill in the south to New Plymouth and Masterton in the North Island, a distance of about 600 miles. The earthquake was severely felt in the vicinity of the Hanmer Plains, in what was then called the Amuri District of North Canterbury. It was followed within the next quarter of an hour by two much smaller shocks, and slight shakes occurred continually until about 6 am. All Saturday, Sunday and Monday, the ground on the Hanmer Plains was constantly quivering.

The government geologist, Alexander McKay, left Wellington for the South Island on 27 September with instructions from Sir James Hector, Director of the Geological Survey, to report on the damage caused.[1,2] His route to the Hanmer Plains took him up the Awatere Valley where, between 29 September and 5 October, he examined the trace of the 1848 fissure, recorded by Sir Charles Lyell and others, that extended along the northern side of the Awatere Valley for 35 miles, from Taylor Pass to Castle River in the upper part of the valley.

Alexander McKay, doyen of New Zealand geology (from the *Cyclopedia of New Zealand* [Wellington Provincial District], Cyclopedia Company, Wellington, 1897). He took the first photographs of the Awatere Fault. After tracing the source of the 1 September 1888 earthquake to movement on the Hope Fault in northern Canterbury and a summer season's geological mapping and collecting in the Awatere Valley in 1888–1889, McKay concluded that the 1848 earthquake fissure was coincident with the line of the Awatere Fault and that it had 'reopened' during the earthquakes.

McKay made inquiries as to the effect of the recent earthquake but found that 'it had excited no alarm'. No damage had 'resulted to buildings or other structures, and no fresh openings along the old rent [the 1848 fissure] or in any part of the Awatere Valley had been made; at all events, none [had] been observed'. At Weld's Hill Station, he learned that the head shepherd had noticed a slip on one of the hillsides, which he supposed to be new and probably caused by the earthquake – the first tangible sign that an earthquake had occurred. At the Jordan Accommodation House, some 5 miles further up the valley, nothing had been damaged although the earthquake had given them 'a pretty good shaking up'. Further on, at Gladstone Station, there was still no physical evidence of the convulsion, although McKay learnt that the effects had been more 'alarming', causing some of the station hands to rush from the cottage expecting that 'something more serious would happen'. A thorough search around the station, on the river flats, and in the adjoining hills, discovered neither 'dislodged rocks nor earth rents'.

From Gladstone to Upcot and on to Langridge Station, McKay was able to again closely observe the line of the 1848 fissure as it was 'not distant more than half a mile, often close to and on the line of the road . . . and for two miles before reaching Castle Creek [Castle River], the road had been made almost along the line of fracture, so

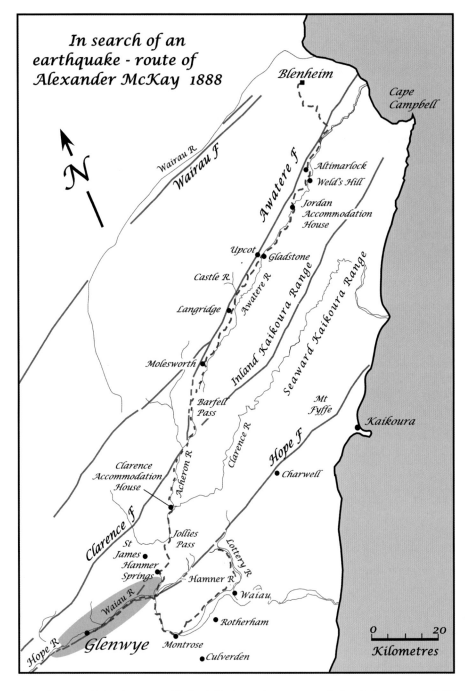

Map showing McKay's route (dashed line) to the source of the 1 September 1888 earthquake near Glenwye in the Hope Valley, North Canterbury, and the surface traces of the Wairau, Awatere, Clarence and Hope faults. The shaded ellipse represents the area of greatest damage during the earthquake.

that there were exceptional opportunities for seeing any fresh displacements that might have taken place. Nothing, however, was observed.'

At Molesworth Station, the information was that the shocks were of no greater violence than at Gladstone and Upcot, and although described as 'severe', they had damaged nothing. For the next 38 miles, across Barefell Pass and down the Guide River to the junction of the Acheron River, McKay saw nobody, and still saw no tell-tale signs of the earthquake. It was not until he arrived at the Clarence Accommodation House that the effect of the earthquake was materially evident from the partly 'ruinous condition of the kitchen chimney of which about 18 inches of the upper part on the side facing the west had fallen. To all appearances it would not have required much of a shock to effect the demolition of the whole chimney but otherwise than as stated it was spared.'

While he was at the accommodation house, McKay learnt that an aftershock on 28 September at 11.30 in the morning had been felt at St James Station in the Upper Clarence Valley with considerable force, second only to the great shock on 1 September. He had also felt this in Blenheim 'as simply a smart shock' and had learnt that there and as far up the Awatere Valley as Molesworth Station it was described as 'a violent shake but not such as to do damage of any kind'. At St James, 'which had been spared the one on the 1st September', the aftershock had done considerable damage and wrecked most of the chimneys. But, as this was far out of his way, he had no time to go there and see for himself.

Reaching Jollie's Pass on 5 October, McKay came across a number of boulders that had fallen from the hills bordering the road. With the keen eye of a geologist, he noted that the stones did not have the appearance of the bleached rock fragments and boulders that were strewn about on the surface of the hills, but were all dark coloured and stained by the soil in 'which their greater bulk have been imbedded'. He concluded that they had literally popped out of the ground during the shaking. At Jollie's Pass Hotel, where McKay halted for lunch, he was told that a little plaster from the older part of the house had fallen, and a loosened slab on the outside of the wall was pointed out to him as having been caused by the earthquake.

The shaking was described as more violent there than to the northeast in the Awatere Valley, and he learnt that at Hanmer Springs, where he was heading, the great shock had shifted a heavy wooden cover weighing more than half a ton over one of the hot springs, and that one of the smaller springs had become so agitated that its waters were thrown over the encasing framework some 6 inches higher than the water level and a good deal of mud was dumped on the surrounding lawn. The ejected mud was still evident when he arrived at Hanmer Springs that evening. Locals described the springs as now 'normal' and certainly not hotter than usual, and said that the escape of gas and mud that had formed numbers of little cones and geysers had been a temporary phenomenon associated with the earthquake's shaking (a process known to us today as liquefaction).

On 7 October, McKay rode south from Hanmer Springs to Mr Low's homestead at St Helens on the east side of the Percival River. Here the shock of 1 September had severely damaged the house, throwing down the chimneys, cracking the walls, dislodging large amounts of plaster, and throwing hams and a birdcage off their hooks. One of the chimneys had fallen through the roof 'causing great havoc in the kitchen'. Another had crashed through the conservatory, while one at the opposite end of the house fell outwards into the garden. The earthquake had caused some pictures to be turned with their faces to the wall, while others, on the same wall, had not been disturbed at all. In the storeroom a heavy fireproof safe was upset and thrown face-downwards on the floor.

While McKay noted particulars of the damage, Mr Low described the 'curious movements of various articles in the dining room'. The dining table stood in the middle of the room, the greatest length of which was oriented in a south-southwest to north-northeasterly direction. On the southeastern side of the dining table was a sideboard with two boxes of plates on the top. The earthquake rotated the sideboard in an anti-clockwise direction, and threw the boxes of plates across the dining table so that they landed on the floor close to the opposite wall – but stacked one on another exactly as they had been before the earthquake. Most oddly, while being thrown across the dining table the lower box must have opened, scattering its plates, while the plates in the upper box were not in the least disturbed.

McKay was also told that some people were thrown out of bed and that one man, trying to escape via the door, had had his course so altered by the motion of the floor that he had run into the fireplace. Although these and other stories were recorded by McKay, he remained somewhat skeptical:

> these stories cannot in all cases be held to be quite trustworthy because I was informed that some of them had been concocted to satisfy the evident longing of newspaper reporters and correspondents for stories of this kind and now there is some difficulty in checking which are genuine and which are not.

The effects of the earthquake were interesting, but McKay was keen to find evidence of ground damage, first-hand information, and observation that he could trust. On the banks of the Percival River, he observed what remained of numerous rents that had opened in the river gravel flats but were now largely filled in. He looked for further evidence of ground fissuring on his way across the Hanmer Plains to the Waiau River bridge, but saw nothing on either side of the main road.

There was general agreement from everyone McKay spoke to that the earthquake had been very severely felt in this part of the plain. From information gathered en route, he learnt that the numerous shocks succeeding the one on 1 September always came from the south, with or without an accompanying noise. These aftershocks affected the surface so that the tall grasses and flax of the swampy parts of the plain indicated the approach and position of the earthquake wave, being shaken as if they were buffeted by a strong gust of wind. When the shocks passed beneath them, people had to hang onto fences or trees to avoid being thrown to the ground by the motion, which was described as 'resembling that which a rat might have, or be subjected to, in the mouth of a terrier'. Tall poplar trees were so moved that they 'appeared as tho' lashed by a perfect hurricane wind and to such an extent that their tops almost if not quite touched the surface of the plain'. At the Waiau Bridge Hotel, McKay was told that a circular chimney of galvanised iron some 12 inches in diameter had been seen to shoot a metre or so into the air in one of the aftershocks and then drop back into its old position.

Proceeding south along the Waiau Gorge, McKay noted three large slips. He had been told to watch out for rocks fallen from the slopes at a place called Marble Point, but looked in vain. At Leslie Hills Station he met the Manager, Duncan Rutherford, who was to be his guide for the next few days. At the station, there was little evidence remaining of earthquake damage. The stone structure of the old house, which had not been 'well constructed to resist even an ordinary earthquake', had been so badly wrecked that it had to be taken down, and at the time McKay arrived a new one was being built of wood.

Mr Rutherford described the shocks of 1 September as very violent. While he had, with difficulty, escaped with his family from the falling building, the men in their cottage close by had not been 'disturbed from their beds and knew nothing of the damage at the Station House till the usual hour for commencing the day's duties'. McKay learnt that the earthquakes, although regarded as the heaviest yet recorded in the district, had done little or no damage at Rotherham and Culverden townships or at Montrose Station. He surmised that the extensive damage done to the station houses at Leslie Hills and St Helens might

> in part be credited to the condition of the buildings wrecked [and] somewhat also to the fact that both stood on alluvial ground although this last as a reason for what happened does not hold good in the face of the fact that Montrose Station, Culverden and Rotherham are also on alluvial plains or river terraces.

Rutherford also told McKay that it had been reported to him that the earthquakes had left more than ordinary traces of their violence on the saddle between the west branch of Lottery River and the Hanmer River. McKay hurried off to see for himself, knowing from his previous geological mapping that this was near the line of an earthquake fault (the Hope Fault) and might show the evidence of ground rupturing that he was looking for. But again, disappointingly, nothing was discernable; no fresh fissures or slips on the hillsides, not even collapsed overhanging creek banks or slips on the edges of the river terraces. And although he was told that a hut at some sheep yards one mile to the north had been thrown down, he did not consider this surprising in view of the condition it was in when he had visited it four years previously.

McKay returned to Leslie Hills Station having satisfied himself that the most easterly effect of the earthquakes was in the Charwell River area near Green Hills, where 'from a steep rocky bank a ton or so has fallen and lies in a pile of rubble at the foot of the cliff', and between Mount Fyffe and Kaikoura township on the east coast, where some chimneys had been damaged 'but not seriously'.

All the evidence told McKay that the centre of the earthquake must lie further to the west. On 9 October, accompanied by Rutherford, he travelled back up the Waiau Gorge to the Hanmer Plains with the intention of visiting Glenwye Station (now 'Glen Wye Station'), near the Hope River. Riding along the south bank of the Waiau River, they were unable to see any effects of the earthquakes until they reached a place called the 'zigzag' where the horse track passed over a bluff some 250 feet above the river washing its base. Beginning their steep descent to the plain, they came to a large east–west fissure slicing across the track. On the river flat at the bottom they found another, and then several more, all of the same orientation, as they rode further upriver. A little further on, where the track left the river bank to cross a high terrace before reaching their destination, they found another fissure, or

> rent opening being 2 to 4 inches and very apparent and here also for the first time is seen the evidences of former earthquakes and like rendings in a line of broken and depressed country not more than one chain in width notching the terrace escarpment and dividing it, here more into two parts, running west along the high terrace.

At last, McKay had found the evidence that he had come all this way to find – the line of rupture and the cause of the first and greatest shock of the 1 September earthquakes. He was now entering the area near the source of the earthquake. Ground disturbance along what turned out to be the 'line of an old fracture' was now very apparent, with numerous surface openings varying in width up to 8 inches, stretching westward in an obvious line.

The road continued over the high terrace, dipping in and out through cuttings formed across the banks of small creeks that ran across it. Where it intersected the line of rupture, every projection on road or creek bank had been sliced off, and this was 'done with exceeding neatness as though made with the exactness of a spade cut but in reality more artistic in execution'. No nook or corner had escaped the slicing action of movement along the fault, and 'it mattered not whither the bank faced the north or any other direction whither forming a convexity, concavity, or angle of whatever kind, the work was done and done completely, and there is not now along the whole line an overhanging bank where a sheep would find the pleasurable occupation of filling its fleece with grit and shade at noon day'.

By now, the bridle track that ran along the south bank of the river was in such an 'utterly wrecked condition' that it became very difficult to follow. The track had been cut partly into solid rock and partly into the overlying river terrace gravels, 'but alike whether or not in rock or shingle the wreck was complete', and at Gorge Creek they were forced to abandon the track altogether and find their way along the higher terraces and slopes south of the line of disturbance as best they could. Here McKay found further evidence that the solid rock had been dislocated by faulting, indicating that the fissures in the ground were not merely superficial features caused by the shaking. At Gorge Creek, where the track was deeply incised through high terraces cut across the bedrock, it had been covered by a large slip which had

Earthquake rupture formed along the Hope Fault at Glenwye, North Canterbury, during the 1 September 1888 earthquake (Alexander McKay, BW4625H, GNS Science).

descended into the river. Beyond, the track continued around the face of the cliff to regain the height of a river terrace some 150 feet above the river. The terrace gravels comprising the upper part of the cliff had collapsed, totally obliterating the track. To 'scramble over and along this with a heavy body of water washing the foot of the cliff was so clearly a risky undertaking' that McKay was forced to retrace his steps and to scale a precipice at another point where he could regain the high terraces.

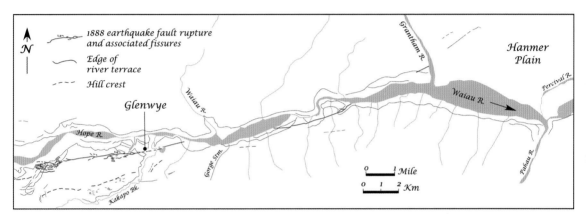

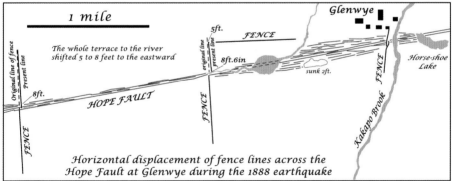

Upper map shows the central part of the 1 September 1888 rupture along what was later termed the Hope Fault, North Canterbury.[3] Alexander McKay's search for the cause of the earthquake, expressed as a surface rupture on a known or new fault line, took him to Glenwye, along the southern side of the Waiau and Hope rivers west of the Hanmer Plains. He followed a horse track that traversed terraced country intersected by deeply incised streams where he discovered an alignment of fissures and associated landslides that represented the effects of the latest movement along an old earthquake fault he had previously recognised further to the east. *Lower map* shows details of the line of ground fissures developed along the Hope Fault at Glenwye Station, and the horizontal displacement (amounts given in feet) of three fence lines that crossed the fault. Redrawn and simplified from the original map by Alexander McKay in 1892 entitled *Sketch Map showing the Effects of the Earthquakes of Sept 1,1888 at Glenwye (Hope Valley)*.[4]

It was now clear that the line of the earthquake's disturbance had run along the Waiau River and across the south bank, cutting obliquely across the high terrace bordering the river. This was evidenced by the presence of fissures and rents, ranging in size from a mere crack up to 10 inches wide and 2 feet deep, in the gravel surface of the terrace. The rents passed through a clump of birch trees near the edge of the terrace, and McKay was surprised to find that all the roots lying across the path of the fissures had been snapped clean through, as though they had been cut with a saw. It was also clear that the line of the rupture was confined between the narrow valley part of the Waiau River and its junction with the Hope River.

Closer to their destination, Glenwye, the ground was extensively fissured and there were numerous slips along a narrow belt of country varying from a quarter of a mile to half a mile in width. Glenwye Station stood on a terrace some hundred feet above the Hope River at its junction with Kakapo Brook. The buildings – a manager's house, the men's cottage, a store, stable, woolshed and sheep dip – were situated close to the foot of the slope up to a higher terrace. The road ascending to the station from Kakapo Brook was in a bad condition and 'no likely place to ride a horse', and the rents and fissures extended in an unbroken line across the lower terrace on which the station stood and continued across the higher terrace.

The men's cottage had been badly shaken – the inside was 'one confused ruin' – and it had been moved north by almost one and a half feet and a little to the east, shifting it from its foundation. At first sight, the Manager's house looked fantastic and rather laughable. It had been twisted and swayed about in various directions, leaving it with a pronounced eastward lean. In contrast, the store, a nearly square building on piles and higher than the other buildings, had suffered little or no damage. The stable also remained intact. It was a comparatively new building and supported by the horse stalls and joists to which the stable posts were fixed. The woolshed was a large building, some 120 feet long and 40 to 50 feet wide, standing with its long axis oriented north–south. It had been severely damaged and presented a 'melancholy appearance' – saddle-backed in two places, bulging inward at the sides, and leaning to the east. All the piles had been tilted 30 to 35 degrees from the vertical, so that the floor swayed to McKay's footsteps as he walked through it.

The cement sheep dip was 'chewed up like a piece of gingerbread in a schoolboy's pocket'. It had been squeezed at ground-level to a little more than half its width, the cement floor cracked along its length in a north–south direction, and the pen rails torn from their housings. Alongside the dip were 300–400-gallon tanks set in masonry, with furnaces below for the preparation of the dip mixture. At the time of the earthquake the tanks were full of water, so each would have weighed about two tons. They had been thrown into the air, and at the same time their masonry base disintegrated so that no two bricks remained bonded together. All that was left was a heap of bricks, on to which the tanks had descended.

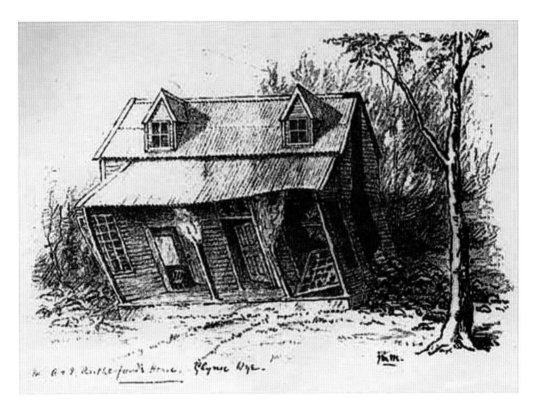

The Manager's house at Glenwye Station, Hope Valley, North Canterbury, damaged by the 1 September 1888 earthquake (*Weekly Press Illustrated Supplement*, 14 September 1888). The brick chimney had totally collapsed. Alexander McKay described the condition of the house and thought that it looked 'fantastic and rather laughable'.[1] In Christchurch, 105 kilometres to the south, the earthquake caused damage to buildings, notably the collapse of the upper part of the cathedral stone spire, and rockfalls onto the Sumner road. In Hokitika and Greymouth, on the West Coast, chimneys were damaged and crockery broken.

One of the fence lines (repaired) at Glenwye Station, Hope Valley, North Canterbury, displaced horizontally across the Hope Fault (extending horizontally across the lower part of the photo as a slight depression containing lighter-coloured stones) when it ruptured during the 1 September 1888 earthquake (Alexander McKay, BW4759, GNS Science).

The manager, a Mr Thompson, told McKay that the 1 September shock appeared to travel 'at a measurable rate' eastwards down the valley, and that it was accompanied by a terrific roaring noise which died away in the distance, while at Glenwye it became momentarily quiet. The aftershocks had caused the buildings to vibrate extremely rapidly in an east–west direction.

McKay noted a large number of funnel-shaped holes in the gravel surface, varying from 4 to 10 feet deep, in the swathe of fissured ground that extended across the terrace flat. He was now to make a simple but momentous discovery. Abreast of the station, a wire fence had been constructed up the slope to the higher terrace and the track passed through it by way of a gate. Before the earthquake, this fence was a straight line, but where it crossed the line of fissures the fence had been strained, the wires broken, and at many places the iron fence poles had been drawn out of the ground.

More remarkably, the fenceline to the north of the fissures was now about 5 feet further east than the fenceline south of the fissures. Again, on the higher terrace some 360 feet above Glenwye, another wire fence crossed the fissured depression. This fence also was 'from end to end straight as put up and remained so until the shocks of the 1st September', subsequent to which it was discovered that the fence had been parted in the depression of the old earthquake rent, and the northern part 'thrown to the east a distance of 8 feet 6 inches' relative to the southern part. The part of the fence that moved eastward was not broken, but the other portion had its wires snapped. Near the line of the ground rupture, the ends of the broken fence line showed a divergence of 4 feet, but at the corner post at the end of the fence line, about half a mile from the rupture, and looking along the line of the fence, McKay judged the lateral shift to be about 5 to 6 feet to the east of where it should have been. Still further to the west, another fence crossed the rupture, and it too had been broken, with the northern part moved to the east by the same distance.

From the high terrace McKay had a good view of the deep, narrow, upper Hope Valley, bounded by high, rugged, bush-clad mountains to the north, and on the south by grass and bare rock bearing stunted bush and scrub. The valley was on the line of the earthquake disturbance, as indicated by large slips carrying 'the fallen debris pell-mell into the gullies lower down the slopes'. Here another kind of earthquake damage could also be seen. Much of the dry standing timber had been snapped off at about 4 metres above the ground by the jerky vibrations. The dry timber had carried with it branches of the living trees as it fell. Along the valley, and the line of the rupture, live trees had also been uprooted – in one instance a tree some 33 feet in height and one foot in diameter, with strong roots fixed into the rocky bank of the river, had been torn from its position and thrown across the track.

McKay had located and followed the line of the earthquake as far as was practical. He had established that the ground damage had been confined to a very narrow belt of country running approximately east–west, beyond which the country to the north and south had been violently shaken; and that the line of rupture followed precisely that of an earthquake fault line that he had mapped previously. He had, for the first time in New Zealand, pinpointed the source area of a large earthquake, or what we would today term the earthquake epicentre. It was clear that no volcanic action had been associated with, or caused the earthquake, as was believed in 1848; it was apparently caused by sudden movement along a section of the Hope Fault. But what force caused the sudden movement that generated the earthquake?

CHAPTER 12

THE AWATERE FISSURE – SOURCE OF THE 1848 EARTHQUAKES

In November 1888, Alexander McKay was back in Marlborough doing summer field work. Armed with first-hand knowledge of the ground disturbance associated with the 1 September earthquake, he made further observations of the fissure along the Awatere Valley. It remained several feet wide, and for part of the way formed a deep ravine that could easily be followed. Everywhere he examined it he was impressed by evidence of recent movements – 'so recent that it may be credited that some of the rents and fractures have been formed within the past forty or fifty years, as testified by the older residents in the district'.[1.] McKay was at first unsure whether the fissure was produced during the great 1855 earthquake that also devastated Wellington, or the 1848 earthquakes, but Mrs Mowat, one of the older residents at Altimarlock Station (through which the fissure extended), confirmed 'that the earthquake of 1855 did not produce any fresh openings along this part of the line; these having been formed by the earthquakes of 1848'.[1] After clarifying the different nature of the rocks on either side of the fissure in several stream sections crossing it, McKay concluded that the fissure was the surface expression of a major fault line, and named it the Awatere Valley Fault (now termed the Awatere Fault). From his observations along this fault and other faults in the Marlborough region, such as the Hope Fault which had moved on 1 September, McKay concluded that

Line of the Awatere Fault and 1848 fissure (indicated by arrow) extending inland up the Awatere Valley from White Bluffs (Lloyd Homer, GNS Science). It was along this section of the fault that 'Earthquake rents' are marked on the British Admiralty Hydrographic map: *Cook Strait*, published in 1878, from surveys made by HMS *Acheron* in 1849.

the earthquakes of greater violence which have taken place since the settlement of the district by Europeans have shown their effects principally along the lines of weakness, producing open fissures, or displacement vertically of the two sides of the fault. Such appearances have usually been regarded as the results of an earthquake or a series of earthquake-shocks, which produced fresh fractures, and first drew attention to the existence of these lines of fracture or earthquake-rents, as they have been called. The fissures and displacements thus produced are usually spoken of as being solely due to the particular disturbance of the time when they were first noticed, but by careful examination of the different lines it is readily ascertained that most of them have only been reopened, and

The prominent line of the Awatere Fault is clearly seen in this oblique aerial photo looking northeast down the Awatere Valley (Lloyd Homer, GNS Science). It was described by the Austrian geologist, Ferdinand von Hochstetter, as a 'remarkable feature', being 'visible on the north side of the river [Awatere River] from White Bluff, over hill and vale for a distance of many miles, right to Berfeld's [Barefell] Pass. In many places, the fissure is said to be 5 to 6ft. deep and 30ft. broad and it resembles a canal without water for miles on end.'[20]

that the latest displacement is but a small fraction of the total resulting from a succession of shocks in times past. A further and closer examination will make evident the fact that no great period of quiescence separates the action last taking place, in historic times, from instances of the same power which have affected the surface at a period prior to the settlement of the country, but which have happened within, say, the last hundred years, more or less. These very modern displacements are easily traced along the surface by the presence of what either is or has been an open fissure, or by the existence of a sudden drop, producing a kind of sunken wall, which latter form can sometimes be traced to great distances. These fissures and displacements occur in mountain-country, in

The prominent trench at ground level marking the line of the Awatere Fault between Lee Brook and Castle Creek in the upper Awatere Valley, as first photographed by Alexander McKay in 1889 (BW4945, GNS Science).

which denudation acts with great rapidity: and if these evidences of earthquake-action and faulting were not of very recent date, or continually being renewed, it is evident that all trace of them would quickly disappear.[3]

McKay made one highly significant suggestion concerning movement on the Awatere Fault. During his survey of the rupture that formed along the Hope Fault causing the earthquake of 1 September 1888, he found that in addition to vertical movement, there had also been horizontal movement, as indicated by the indisputable evidence of the displaced fence lines. In what must have been a flash of inspiration, he saw that this small-scale horizontal movement might have relevance to long-term movement on the Awatere Fault. And the evidence for this came from his matching up of similar rocks on either side of the fault which indicated that they must have been horizontally displaced by some 15 miles! Such a large displacement represented the sum of incremental movements along the fault during many past earthquakes over a long time.[2]

The surface trace of the 1848 earthquake 'crack', seen today near the Redwood Pass road as a line of wet ground marking the largely infilled fissure.

McKay's imaginative leap from a small-scale single earthquake movement to large-scale long-term displacement on faults was an important development in the science of seismology. Unfortunately, at the time, this novel and far-fetched notion was considered highly improbable, the geological evidence was not precise, and it was deleted from the published version of his original report dated 1 January 1890, written from Gladstone Station in the upper Awatere Valley, to his superior, Sir James Hector.[3] McKay's idea was promptly forgotten for the next 66 years of geological work in New Zealand. It was only in the mid-1950s that New Zealand geologists began to recognise that movement on all the major faults that they had mapped was predominantly horizontal rather than vertical. Such is the nature of scientific progress.

Although McKay had recognised the possibility of large-scale horizontal displacement of rocks on either side of the Awatere Fault, he completely missed the small-scale field evidence of horizontal displacement that had occurred on the fault during the first great shock of the 1848 earthquakes. In 1848, there were no

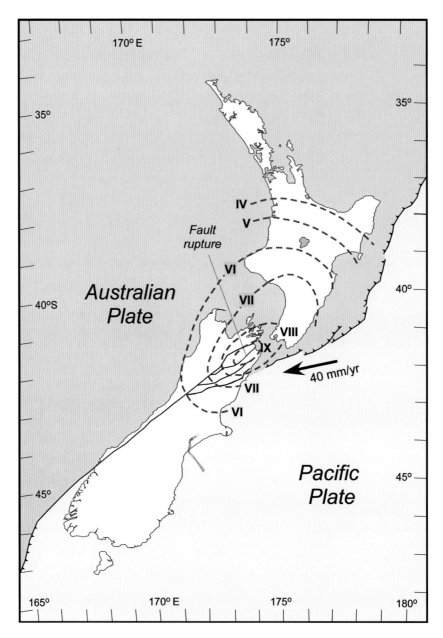

Felt intensity map of the 16 October 1848 earthquake (estimated magnitude of 7.5) constructed from reported damage in different parts of New Zealand (after Fig. 7 of Grapes et al. 1998),[7] together with ellipses of equal earthquake intensity (known as isoseismal lines and designated by Roman numerals) according to the New Zealand Modified Mercalli (MM) scale (see Appendix 2). The area of maximum shaking intensity is centred along the Awatere Fault rupture that caused the earthquake. The Pacific and Australian plates are converging in the Marlborough–North Canterbury area at about 40 millimetres per year and this causes large earthquakes in this part of New Zealand.

man-made features, such as fences built across the line of the fissure, to indicate the amount and nature of movement along the fault. To the eye of even as experienced an observer as McKay there was nothing obvious to indicate what, if any, movement on either side of the line of rupture had taken place in 1848, except for the formation of a fissure in the ground.

However, McKay's realisation that the line of this 1848 fissure was coincident with what is now called the Awatere Fault was important.[4,5] This fault is now known to be one of the major earthquake fault lines that extend through the Marlborough district and comprise a wide zone of deformation where the Australian Plate to the west converges with the Pacific Plate to the east. From the air and the ground, the Awatere Fault is a remarkable feature – a distinct knife-like cut through the landscape, as recognised by earlier observers, which extends some 175 kilometres from Cook Strait to the Southern Alps. The distinctness of the fault line has been enhanced by repeated earthquake rupturing over several thousand years and the low rainfall conditions of the Marlborough area, both ensuring long-term preservation of the surface trace of the fault.

In 1848, however, only a surface rupture or fissure was seen. Today, the infilled 1848 fissure can still be recognised in several places as a strip of greener grass and rushes, indicating slightly wetter conditions than normal. Landform features that provide evidence of the smallest displacement – such as dog-legged streams and abandoned stream channels, or the disrupted edges and flats of river terraces – indicate that both the southeast and northwest sides of the fault have been uplifted between 0.4 and 3.5 metres, and that there has been a larger horizontal displacement of between 4 and 7 metres, with land on the north side of the fault moving northeastwards relative to land on the south side of the fault [6,7,8,9] – the same sense of horizontal movement recognised by McKay from large-scale displacement of rock formations, and the displaced fence lines crossing the Hope Fault at Glenwye. These smallest displacements are inferred to represent the movement that occurred on the Awatere Fault when it last ruptured, producing the earthquake early in the morning of 16 October 1848. Such small-scale displaced features along the fault have now been recognised for a distance of about 110 kilometres, from the coast at White Bluffs to Barefell Pass in the upper Awatere

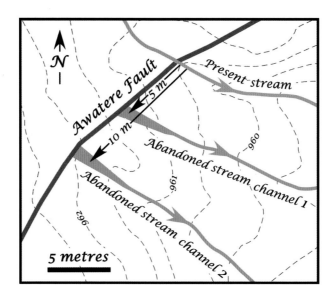

Small-scale map showing horizontally displaced stream channels crossing the Awatere Fault (simplified from Figure 3 of Mason & Little, 2006[9] – the surveyed site is close to Molesworth Station in the upper Awatere Valley, about 92 kilometres from the coast at White Bluffs). The abandoned stream channels (depressions with no water flow) have been shifted laterally from the position of the present day stream by earthquake movement along the Awatere Fault. The last earthquake in 1848 moved the first abandoned stream channel by 5 metres; the second abandoned stream channel was also moved by 5 metres in the 1848 earthquake and another 5 metres during a previous earthquake (10 metres in total). The lateral displacement of the second stream channel thus records the combined movements of two earthquake ruptures of approximately equal magnitude (estimated at about 7.5 for the 1848 earthquake) on the Awatere Fault. The next earthquake produced by rupturing on the Awatere Fault is expected to displace the present stream channel in the same way. After the earthquake, the stream will readjust itself to a new course across the fault line and there will be three abandoned channels on the southeast side of the fault to record the three separate earthquake displacements.

Valley.[7,9] The data, from analogy with modern determinations of average horizontal displacement relative to the length of the fault rupture (in this case at least 110 kilometres), imply that the 16 October 1848 earthquake probably had a magnitude of between 7.4 and 7.7.[7,9] By comparison, using the same criteria, the magnitude of the 1888 North Canterbury earthquake is estimated to have been 7.0–7.3, with rupturing of the Hope Fault over a distance of about 30 kilometres.[10]

The landform reference lines and surfaces that provide evidence of horizontal and vertical movement along the Awatere Fault are also associated with all the major earthquake faults in New Zealand. In the case of the Awatere Fault, they prove a

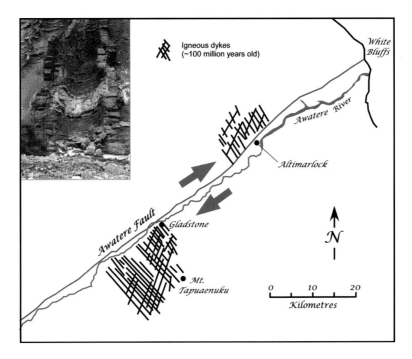

Map showing horizontally displaced clusters of igneous dykes along the Awatere Fault determined by Alexander McKay in 1892. The dykes are intrusions of magma up near vertical fissures in sedimentary rocks (as shown in the inset photo) that were formed between 100 and 60 million years ago. The dyke clusters on either side of the Awatere Fault are separated by about 34 kilometres and they can be used as markers to provide an estimate of the total amount of lateral movement that has occurred on the Awatere Fault as a result of many earthquakes (possibly several thousand!) since the time the dykes were formed. The distribution of the dyke rocks is taken from McKay's coloured map entitled *Geological sketch map of the provincial district of Marlborough and of S.E. Nelson.*[4] As McKay's map does not show the Awatere Fault, it has been drawn on this map.

history of repeated earthquake rupturing and movement going back at least 65,000 years,[11,12] the youngest and smallest displaced features being those formed during the latest earthquake in 1848. Radiocarbon dating of wood fragments in gravel and mud layers exposed in excavations at several places across the fault, suggest that nine large earthquakes (including the one in 1848) have ruptured the fault over the last 8600 years, five occurring between 5600 years ago and 1848. The data also implies that the earthquakes are not of regular occurrence, with 'return times' of anywhere between 40–640 and 1350–2300 years, and indicates that the earthquake that ruptured the Awatere Fault before 1848 occurred between 980 and 1080 years ago.[11,12]

It is the convergence of the two vast rock masses that form the Australian and Pacific plates, with the Pacific Plate sliding underneath the Australian Plate at a rate of about 40 millimetres per year, that explains the cause of large earthquakes. In Marlborough and North Canterbury, where the plate boundary zone swings inland from the Pacific Ocean, it is some 150 kilometres wide. In addition to well-developed active faults such as the Wairau, Awatere, Clarence and Hope faults, there are dozens

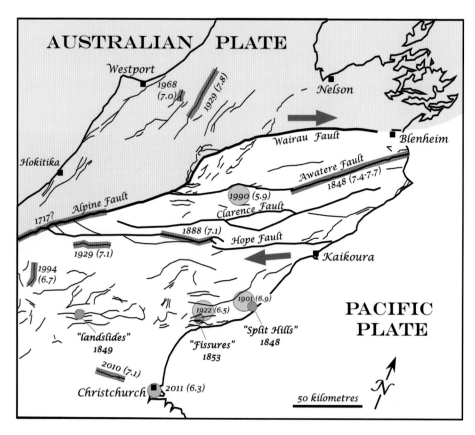

Map of the northeastern part of the South Island showing the major fault lines (thick black lines, named) that form the boundary zone (white) of the Pacific and Australian plates. Other faults (thin black lines, unnamed) are also shown, together with fault ruptures formed by large earthquakes (red lines, with year and magnitude). The epicentres of other earthquakes that did not cause surface faulting (1901, 1922, 1990, 2011) are also indicated (pink circles). The three areas with earthquake ground damage that may have occurred during the 1848 earthquakes, described between 1848 and 1853, are indicated by small grey circles. The thick grey arrows indicate the direction of horizontal movement on opposite sides of the boundary zone: to the east on the northern side of all of the faults shown, and to the west on the southern side of the faults.

of smaller ones, each with the potential to rupture and cause a large earthquake. While many faults are known from surface features, others, such as the one that caused a 7.1-magnitude earthquake on 4 September 2010, badly damaging the Christchurch area, are hidden – in this case, buried beneath 16,000-year-old river gravels. As the plates continue their inexorable convergence, the great thicknesses of rock that compose them undergo a build-up of strain and they slowly deform. Some of this strain may be released by small, non-damaging earthquakes, most of which (and there are thousands per year) are only registered by seismographs. But if the strain accumulates beyond the strength of the rocks, such as when the interface between the plates becomes periodically 'jammed', they will suddenly break, releasing a huge amount of energy in the form of a large earthquake, either as movement along a pre-existing fault or by the creation of a new fault.

If the first great shock at 1.40 am on 16 October 1848 was caused by movement on the Awatere Fault, what caused the other shocks that were reckoned by Wellington's inhabitants to be of equal or greater intensity? Could the first great shock have triggered ruptures along other faults? A report in the *New Zealand Spectator and Cook's Strait Guardian* of 25 October 1848 about a 'remarkable' fissure extending across the beach at Thorndon is interesting in this respect because its location coincides with the surface trace of the Wellington Fault.[13] Perhaps movement on the Wellington Fault caused the shock on the afternoon of Tuesday 17 October, or possibly the one that occurred early in the morning of Thursday 19 October? Unfortunately, the presence of the fissure was only reported after all the large shocks had occurred, and no further information is available.

An account by the HMS *Acheron*'s naturalist, Frederick Strange, who observed large slips inferred to have been caused by the 1848 earthquakes in Canterbury, is also intriguing. During a 60-mile journey from Port Cooper (Lyttelton) inland to the Snowy Mountains (Torlesse Range) early in 1849, he reached the summit of a 'high snowy peak', and there

> the sight that met my view was very singular and wild: whole sides of mountains
> appeared to have slipped into immense gullies below, whilst immense blocks of
> rock had been precipitated, cutting their way through black-birch trees which

line the gullies and carrying everything with them. The disruptions of many of these rocks appeared to be very recent from the appearance of some of the trees that had been snapped. I feel confident that these traces of recent convulsions on the mountains are the effects of the late earthquake which did so much damage at Wellington.[14]

Warren Adams found further evidence of ground disturbance, probably attributable to the 1848 earthquakes, some 90 kilometres to the northeast along a range of limestone that forms the eastern slope of Mount Cass. He described a prominent terrace about 'twenty yards in width . . . much broken by deep cracks, occasioned by earthquakes, and extending for an unknown depth.'[15] About 45 kilometres further northeast towards the coast, the hills between Port Robinson and Cheviot were said to have 'been split and one side subsided in places for five to six miles' after the earthquakes.[16]

The landslides described by Strange are close to the surface trace of another active fault, the Porter's Pass Fault, and the locations of the other earthquake disturbances are situated on a similar northeast trend, all of which hints at the possibility that this fault or other collinear faults may also have moved, causing one or more of the 1848 aftershocks felt in Wellington* – or possibly the shock felt by Frederick Weld and Thomas Arnold on Sunday 22 October at Flaxbourne, which rocked the hill on which they were sitting,[18] and was recorded in Wellington as a 'slight shock' at 10 am.[19] But all this is mere speculation, and in the absence of further definitive evidence, is likely to remain so. Nevertheless, the occurrence of three large earthquakes in October 1848, closely spaced in time, with the first not necessarily the main shock in terms of strength, is an unusual occurrence in New Zealand's historical earthquake record.

---

* Radiocarbon dating of wood fragments from probable earthquake-induced landslides in the Torlesse Range close to and covering the Porter's Pass Fault indicates ages of between 500 and 700 years before present, and therefore older than 1848. These are probably the same landslides observed by Frederick Strange. It is possible that some of the old landslide scars may have been reactivated by earthquake shaking in 1848 and are not related to 1848 movement on the Porter's Pass Fault.[17]

CHAPTER 13

FUTURE VISITATION – AN IMAGINARY SCENARIO OF A LARGE EARTHQUAKE AFFECTING WELLINGTON

Wellington,
16 October 2037

By now you have no doubt heard about the devastating earthquake that has affected the central part of New Zealand, and in particular Wellington. I'm alright but as you can imagine, things are far from normal. It feels as though I'm living on jelly! The movement of the ground has been incessant from all the aftershocks, two of which have been strong enough to cause further damage to buildings that had already been weakened and stressed by the first quake that struck at 8.30 this morning.

At that time I was just coming out the door on my way to work and walking to the garage when, without any warning, the shaking began. It intensified so quickly that I was thrown onto the lawn. Trees and the surrounding houses were vibrating at such a rate that I could barely focus on anything, and the ground motion was such that I began to feel sick. The noise was terrific, a rumbling, roaring sound that seemed to be everywhere, accompanied by the combined sound of creaking, grating and tearing. I remember seeing the cat shoot past me to goodness knows where. Inside, the unforgettable sound of breaking glass, the smash of crockery and the thudding of falling furniture. It must have been nearly 30 seconds before the shaking subsided. I lay there on the lawn, motionless, my mind still not having caught up with the experience I'd just been though. It was strangely quiet although the ground was still quivering. I began to look around. Neighbours were beginning to emerge from their houses. They all had blank looks. At first I couldn't see any building damage. The houses looked the same. Chimneys were still standing. The front windows seemed intact. But then the effects of the shaking came into focus. The chimney next door had a diagonal crack down its length. Across the road, a chimney pot had fallen off. My chimney had a wide crack around the top. Two of the bedroom windows had cracked. One of the neighbour's windows had popped out. Water was running down the road and from the front gate I could see a fountain of water spurting upwards from a crack in the road near the corner. Undoubtedly the water mains had broken. I wondered about the underground gas pipes and electricity cables. Then I remembered the house – what was the inside like? I had to go and look. From the front door, I saw that all the pictures in the hall had been catapulted off the walls. The kitchen was chaotic. Plates, cups, bowls and glasses were all over the floor. Most had smashed. The doors of the cupboards from which they were thrown were still swinging, as were the hanging lamps. Even the refrigerator door had burst open, spilling most of its contents over the floor. In the dining room, my glass cabinet full of all the semi-antiques I had collected over the years was lying flat on the floor. Through the glass side panels I could see that most of the things inside were broken. Going though to the living room, vases and papers on the sideboard had been thrown onto the floor along with the stereo and speakers. The television was face-down on the floor. A couple of pictures had fallen and the rest were turned sideways. In the library, nearly all of the 900 or so books had ended up in a large heap on the floor and desk, although the book cases that had been fixed to the wall remained upright. But the large window in the room had broken, half of it falling outwards into the back garden. The plaster ceiling had been cracked from one corner to the other and chunks of plaster had also come down. It was difficult to get into the music room because the double doors had jammed, and in trying to force them open, I broke one of the glass panels. Inside, apart from a few loose music scripts and the metronome that were on the floor, no damage had been done. Amazingly, the piano had been shifted into the middle of the room and slewed around. The bathroom was a bit of a mess: toiletries had spilled out of the cupboard and the mirror had fallen into the basin and smashed. I turned on the cold water tap – only a trickle of water; flicked the light switch down – nothing: power and water supply gone. The telephone line was dead. Looking out the kitchen window down to part of the Karori valley, the air was hazy with dust. Traffic had stopped and I could see people moving about. There didn't appear to have been any accidents. I could see that some chimneys had fallen but there was no smoke that might indicate that fires had broken out. Some of the overhead power lines were down and I saw that a few of the posts had fallen, apparently snapped off near their bases. There was no sound of any sirens. Another violent shake sent me back under a door frame and then another sent a tottering glass or plate crashing to the floor. Once again outside, I thought about the cat but it was nowhere to be seen. It was probably hiding under the house, but I didn't feel like going into the basement to have a look with all the shaking going on.

So what now? I started to assess the situation. The house did not appear to have suffered any major structural damage. The walls were still vertical, as was the chimney, and the roof still attached. Thank goodness for wooden houses! A few windows had gone. A good deal of mess inside and a lot of breakage, and no future joy in cleaning that lot up. Power, water and telephone cut – what about the food in the freezer? And how much water have I got? Water in the toilet cistern, a good amount in the hot water cylinder, and in the header tank for the shower, some bottles of fruit juice and milk from the refrigerator, now on the floor. I have a barbecue and a full gas tank and enough wood for a fire if it comes to that, so cooking and boiling water should be all right. I switched on the transistor radio but got nothing. By this time people up and down the street had started talking, checking on and comforting one another, assessing the damage. The old lady two doors away had fallen and badly bruised her leg. A few mothers were desperate to get to the local primary school to check on their children. There was a lot of cellphone activity. Unfortunately, I hadn't charged my phone battery for some time and so that proved useless. Nobody rang me. At least I was unhurt. It must be awful not to know about those who had left home before the earthquake.

The 'big one' that everyone had expected, the one that we had been told would come sometime, had finally arrived. Having got through the first terrifying seconds of the initial shock we had yet to go through the aftershocks, the growing anxiety, and the gradual realisation of how widespread the devastation would be. At this stage nobody knew anything except what had happened in part of our own street and in a little part of one suburb of greater Wellington. How had other places fared? What about in the city itself? It was so frustrating not knowing, but even if we did know, what could we do about it? I decided to get the ladder and binoculars and go up on the low pitched iron roof of the house for a better look. Perhaps a silly idea in view of the possibility of strong aftershocks, and remembering the experience of poor old John Plimmer during the second great shock of the 1848 earthquakes, but I decided to go anyway. Up on the roof I had a good view over most of the suburb and I could also see part of the main arterial route in and out of Wellington along the western side of the harbour. Through a faint haze of rising dust, the morning sun was glinting off almost unbroken lines of cars. They were not moving and there had been three multiple car pileups, two in the incoming lane and one in the outgoing lane. The roadside strip along which the railway ran appeared to have slumped outward into the harbour and I could clearly see the wavy outline of the rail tracks. But more ominous were two landslides that had come down from the steep hills on the other side of the road. Both landslips had crossed the road and gone into the harbour, and they had undoubtedly buried some cars. I could clearly see the back of a car sticking out from the rubble of one of the landslides. Most drivers were out of their cars. Many were crowded around the accident sites, others were on the landslide debris, looking for survivors, or at least that is what I reckoned. Then I noticed the smoke and dust rising rapidly upwards in the almost still air from the city area hidden behind the eastern hills of Karori. I wondered if the smoke was from one or several fires – perhaps ignited gas from broken gas mains. It was only now that I could hear sirens, also coming from the direction of the city. I tried the radio again, but still nothing. In the valley below there had been more damage than I had first seen. A lot more chimneys were down and a number of roofs damaged. A few houses were on a lean, most likely due to foundation failure that caused some of the piles to sink. Traffic had been stopped by fissures across the road in one place. Areas of water and mud indicated the failure of water pipes, or perhaps liquefaction. There were a lot of people walking around, many having got out of the stationary buses and presumably on their way home. Looking around at the surrounding hills I saw that there had been a few slips that had taken out vegetation, but nothing of any significant size. Dust was rising from the steep western slopes of the Rimutaka Range, indicating that larger landslides had occurred over there. Having seen all there was to be seen, I was climbing down when another sudden shock caused the ladder to move sideways along the guttering. I instinctively leaned in the opposite direction to prevent it moving any further, and failing to do this jumped backwards and landed on the lawn. Luckily, I had been more than halfway down at the time of jumping and so escaped injury except for a bruised thigh. I was not going to attempt that foolish exercise again.

It's fortunate that the house is built on rock, although, like so much of the Wellington hills, it is rotten rock, covered with a thin veneer of soil. Nevertheless, this relatively solid foundation spared the house and those around from any major damage. Those sited on alluvium in the Karori valley have clearly sustained more. I thought of all the buildings situated on relatively soft, sediment-filled valley and low-lying areas in the Wellington region – the reclaimed strip of land fronting Thorndon, around Lambton Harbour and Te Aro in the city, Newtown, Rongotai Isthmus, Miramar and Seatoun to the east, and the Hutt Valley north of the harbour – where amplification of the intensity of ground motion, as in Karori, would be expected to occur in a large earthquake, and wondered how they had fared. It would be interesting to see if this earthquake vindicated the ground-shaking hazard-zoning map of the Wellington area completed a few years ago.[1] I wondered also how the residence of the British ambassador, the site of Judge Chapman's house 'Homewood' at the time of the 1848 earthquakes, and only a few kilometres from where I live, had come through the shaking.

It was time to try the radio again, and this time I was able to hear some news. The earthquake had registered a magnitude of 7.6 on the Richter scale, originating

Oblique aerial view looking southwest across Wellington Harbour and city, and the Wellington peninsula. The eroded scarp of the Wellington Fault is an obvious linear feature cutting through the landscape (the actual surface trace of the fault is indicated by the dashed white line). Other localities mentioned in the text (including that of the author's house) are also indicated. In the distance beyond Cook Strait is the Marlborough area of the South Island.

at a depth of about 30 kilometres and located about 40 kilometres southwest of Wellington beneath Cook Strait. Reports were coming in of damage in Nelson, Blenheim and Kaikoura to the south, in the Wairarapa to the east, in Wanganui to the north and in all the towns from there towards Wellington. Wellington had suffered considerable damage, as had the Hutt Valley, Porirua and Wainuiomata. The earthquake had triggered slips in the Ngauranga Gorge, on State Highway 1 along the Paekakariki coast, and from the hills bordering the motorway along the western Hutt Valley, as well as quite a few from the eastern hills, and on the road over the Rimutaka Range to the Wairarapa. At several places rock falls had blocked the coastal road both inside and outside the harbour. The road bridge over the Hutt River at Silverstream had collapsed, as had the railway bridge further down the river. Most of the information was coming in from helicopter observation relayed to the local radio station. I could see and hear several helicopters buzzing around.

Apparently, Wellington Airport was out of action, as the main runway had a couple of cracks stretching across it and there had been subsidence at its southern end. All incoming flights had been diverted.

There had been a number of people killed and injured, although as yet no confirmation of numbers. Many of the casualties had apparently resulted from people being hit and cut by splintered glass falling from the tall buildings along Lambton Quay and the northern end of The Terrace. Glass, a lot of paper, and some heavier objects, such as potted plants that had been standing near the windows, had fallen from many of the high-rise buildings onto the streets below. The radio warned that with the continuing aftershocks, falling glass was still a major hazard (and also watch out for the pot plants, I thought). Ruptured water pipes had caused internal flooding of several floors in some of the buildings. Collapsing parapets along shop fronts, particularly in Manners and Willis streets, along Lambton Quay, and in streets towards the harbour, had also resulted in casualties and had crushed a number of parked cars. It appeared that most of the obvious damage had been sustained by buildings along or near the junction between the old shoreline and the belt of reclaimed land fronting Lambton Harbour. The Te Aro area seemed to have suffered most, just as it had in 1848 and during subsequent damaging earthquakes.

Wellington was temporarily paralysed. As the quake had occurred at the peak time of the morning rush, the inner city roads had become gridlocked. Most people had left their vehicles, public transport, offices and shops and were either congregating in open spaces or streaming out of the city on foot and bicycles, heading for their homes. Understandably, there was considerable confusion and panic as people tried to find their friends and relatives, and search for those thought to be trapped in buildings by jammed electronic doors and lifts. Ambulances could not travel very far and many of those injured had to be carried by hand to them or to places where helicopters were able to land. Water pipes had burst and in at least three places gas mains were reported to be ruptured. Gas from one of these ruptures had ignited and was still burning.

I continued listening with mounting alarm to updates of the destruction caused by the earthquake for the remainder of the morning, and then decided to do something about the mess inside the house. There was still no electricity and no telephone. Water continued to flow down the street. According to one of the neighbours who had a gas connection, there was no gas supply either. I simply dumped the broken crockery, glass and spilt food into plastic bags in the back courtyard and tidied things up as much as possible. I imagined what all the supermarkets and liquor outlets, with everything arranged neatly but precariously right up to the edges of the shelves, must look like now. That would be an interesting cocktail to taste! Everything that looked as if it would fall I took down and stacked on the floor. The fallen books could be left for the time being. It was not at all pleasant doing all this with the shaking still going on. There were some anxious moments as now and then the background tremor was punctuated by a short, sharp shock, sometimes by a longer, swaying one. They seemed to come from various directions and were usually accompanied by a low rumbling sound. I ended the morning with a 'picnic' lunch on the front lawn.

The latest update from the radio said that 236 people in the Wellington–Hutt Valley area had been confirmed dead and over twice that number had been injured. What would the final figures be – and what about places north of Wellington, and in the South Island? First aid centres had been set up in open spaces and the more seriously injured had been taken to Wellington Hospital, which, although it too had been damaged, was in full operation, using emergency generators.

About 20 minutes after the earthquake, the water in Lambton Harbour, which was nearing high tide level, suddenly rose and flooded the Lambton Quay area. It then withdrew again to its normal level. It was thought that this might have been the effect of a tsunami generated by the earthquake, because the same rise, accompanied by waves, had come over the road along the south coast and along the Eastern Bays on the other side of the harbour. The interisland ferry reported experiencing a sudden generation of waves in the middle of Cook Strait on its way over to Wellington from the South Island, and those on board clearly felt the earthquake shock as a distinct jolt that made the whole ship shiver. Pilots of an airliner coming in to land at Wellington Airport at the time of the earthquake had the unique experience of clearly seeing not only the runway, but all the Wellington hills in motion, and had only a few seconds' warning from the control tower to abandon the landing and return to Christchurch.

By late morning Civil Defence measures had all kicked into action. Power to some areas had been restored. Attempts were underway with heavy earthmoving machinery to clear the roads blocked by landslides and rock falls, although the effort was hampered by all the abandoned vehicles. Fortunately, many had left their car keys still in the ignition, but a considerable number of vehicles had to be bodily shifted to clear the way. There were numerous cracks and signs of liquefaction reported in the area of reclaimed land fronting Lambton Harbour, particularly in the area of the Container Terminal. Seaward edges of the terminal and other places had slumped, as had many of the wharf piles along the harbourside quays. The colour of Wellington Harbour had changed to a greyish brown, and a growing slick of oil indicated that the diesel fuel pipeline between Aotea Quay and the bulk storage at Kaiwharawhara had ruptured. Reports were also coming in that oil was leaking from the big storage tanks situated on reclaimed land at Seaview, in the northeastern part of Wellington Harbour.

At around two o'clock I heard that the Wellington Fault had ruptured. Observation from the air showed a clear swathe of landslides, broken trees and ground fissures right along the fault line from Cook Strait to at least Kaitoke, a distance of about 75 kilometres. The Karori reservoir, through which the fault passes, was still intact, but the main access route to Karori, the Karori Tunnel, was blocked by rock falls at both ends, as was the road over the hill. Where the fault entered the harbour, the vehicle access ramp to the interisland ferry had collapsed, and the vehicle marshalling area, on reclaimed land, was badly cracked and had shifted outwards into the harbour. The highway entering Wellington had fortunately withstood the shock except for a few minor cracks, vindicating the use of lead-rubber earthquake shock absorbers in its construction. Along the western side of the harbour, where the fault runs offshore, people trapped on the highway had seen a long line of furiously bubbling water, caused by the escape of gas at the time of the earthquake. The water had quickly become turbid and changed to a brown colour from disturbance of the bottom sediments. All along the western side of the Hutt Valley there were extensive fissures that had formed parallel to the line of the fault and from which liquefied mud and sand had been ejected, together with water. Fissures and holes from which water and sediment had been expelled had appeared over large areas of Lower and Upper Hutt, and were especially numerous in the Petone area in the lower part of the valley. Liquefaction had also caused part of the wall fronting Petone Beach to subside, although the 400-metre-long wharf appeared to be still intact (apart from its southern end, which was tilted downwards). Numerous fissures had opened along the banks of the Hutt River. The western side of the fault had been uplifted relative to the eastern side, the uplift appearing as an almost continuous vertical step up to a metre high in the Petone–Lower Hutt area. That probably meant that the land on the eastern side of the fault, including Wellington City, had dropped. No wonder the sudden rise in the water level of the harbour was able to flood a large part of the central city. (This change in elevation is going to be a major problem to deal with in the future, I thought.) The information at hand indicated that, in addition to vertical movement, offset curbs and the central white lines of roads had been moved apart horizontally by 3 to 4 metres where they crossed the fault line, those on the western side of the fault moving northeast relative to their continuation on the eastern side, which had moved southwest. That amount of shift must also mean that all water, sewerage and gas pipes that crossed the fault line must be likewise broken and displaced; and it probably explains why the rail and road bridges further up the valley collapsed – because they too had been built across the fault. In a day or so, the Wellington geologists will be all over the fault line, mapping, photographing and measuring down the last millimetre. What an opportunity! There is no doubt that from the scientific, technical and sociological points of view, the earthquake will provide material for an enormous amount of research that is going to keep people busy for the next few years. Part of the aftermath will be the ruined capital of New Zealand becoming the Mecca for all kinds of experts, who will visit to document in detail yet another of the world's earthquake disasters – this time it's ours. Of course, that's going to be small comfort to those that have suffered.

Luckily, the weather has remained fine. I managed a barbecue dinner with neighbours, but it was a gloomy gathering, the radio continuing to tell us what we would rather not hear. We felt for all those families that had lost loved ones, and for those injured, having to endure pain and discomfort while the rumbling, swaying and jolting continued. We all desperately wanted to know more, but that would come tomorrow and during the following days. Everybody was tense and anxious. Tiredness was evident on everyone's faces. I thought about those Wellingtonians who had gone through three large shocks in the space of only a week during the 1848 earthquakes. They were a tough lot back then. I just hope we are not in for the same. If we are, I don't know if we could stand it. But for now, there is nowhere else to go and there would be no point in going anyway. As night began to draw in, a northwesterly sprang up. The pre-earthquake early morning forecast had predicted rain in the evening and that decided where I was going to sleep – inside the house. For safety I decided to move a mattress under a table and sleep there, in case the ceiling plaster, or even the ceiling itself, came down in a large aftershock. Hopefully, if that happens the floor will hold up.

It's now ten o'clock in the evening. I've been typing for about two hours and have now just realised how totally exhausted I am from all the stress and anxiety of the day's events. It must be the adrenaline that's keeping me going, although it probably does me some good writing about it. I'll continue telling you about developments tomorrow until the phone line is restored and I can send this e-mail off to you. Everyone is going to have a story to tell. Another sharp shock lasting a few seconds has just finished. I just hope we all get through the night safely. Despite a thorough search and repeated calling, the cat's still missing . . .

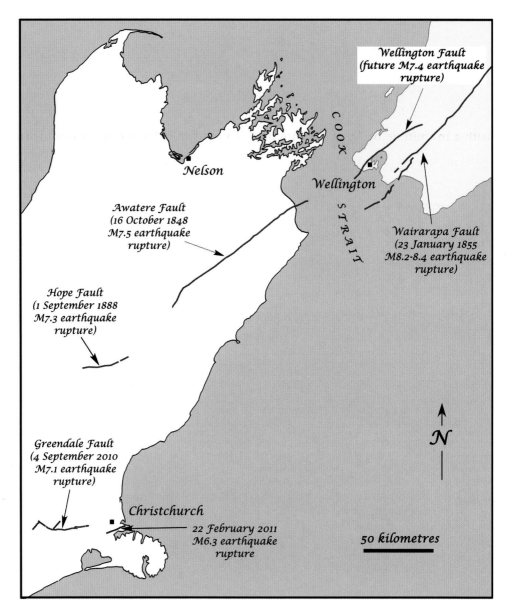

Map showing a possible future earthquake rupture (about 75 kilometres long) along the Wellington Fault (from Kaitoke to 20 kilometres into Cook Strait) generating an earthquake of possibly 7.4–7.5 magnitude, compared with fault ruptures formed during earthquakes in 1848 (Awatere Fault), 1855 (Wairarapa Fault), 1888 (Hope Fault), 2010 (Greendale Fault), and 2011 (unnamed fault). Data for the 4 September 2010 and 22 February 2011 earthquakes are from the GeoNet site of GNS Science. Red fault lines are surface ruptures; blue fault lines are subsurface ruptures.

Postscript

Since the above scenario of a future Wellington earthquake was written, Christchurch has been extensively damaged by two earthquakes – the first with a magnitude of 7.0 at 4.35 am on 4 September 2010, and the second with a magnitude of 6.3 at 1.51 pm on 22 February 2011. Both were unexpected. The 4 September earthquake occurred 10 kilometres beneath the Canterbury Plains, 20 kilometres from Christchurch, by rupturing of a fault buried beneath undisturbed river gravels that were 16,000 years old. A surface rupture extending for 21 kilometres was produced. The 22 February earthquake was caused by sudden movement on another hidden fault close to Lyttelton. This time the focus of the earthquake was only 5 kilometres deep and the shaking effects were much more severe. In contrast, the Wellington Fault is an obvious feature in the landscape and is substantially larger. A possible rupture along the fault during a large earthquake of 7.4–7.5 magnitude could be 75 kilometres long, extending from beneath Cook Strait to Kaitoke. Radiocarbon dating suggests that the last earthquake generated by movement on this segment of the Wellington Fault occurred sometime between 1640 and 1840 AD.[2] Many of the effects of the Christchurch earthquakes, such as liquefaction of low-lying areas underlain by sand, silt and gravel, and landsliding from steep slopes and cliffs, would also occur during a strong earthquake in Wellington. But other effects, such as seiching (the sloshing of water) in Wellington Harbour, and the possibility of a tsunami being generated in Cook Strait, affecting the coastline and Rongotai Isthmus, where the airport is situated, would be unique to the Wellington region. Urbanisation and infrastructure development in the Wellington region, particularly over the last 150 years, has significantly changed the landscape since the earthquakes of October 1848, making it much more vulnerable to the damaging effects of strong shaking during a large earthquake.

REFERENCES

ATL = Alexander Turnbull Library, Wellington
NLNZ = National Library of New Zealand, Wellington

Preface

1. *New Zealand Gazette*, 30 May 1840.

Introduction

1. Grapes, Rodney. 2000. *Magnitude Eight Plus*. Wellington: Victoria University Press.

Acknowledgements

1. Eiby, G.A. 1980. *The Marlborough earthquakes of 1848*. DSIR Bulletin 225, Wellington: New Zealand Government Printer.
2. Grapes, R., Downes, G. & Goh, A. 2003. *Historical documents relating to the 1848 Marlborough earthquakes*. Institute of Geological and Nuclear Science report 2003/34.

Chapter 1 Wrecked

1. *New Zealand Spectator and Cook's Strait Guardian*, 25 October 1848.
2. Lt. Gov. John Eyre to Governor George Grey, despatch, 29 October 1848, in *Papers relative to the Recent Earthquake at Wellington. Documents presented to both Houses of Parliament of Her Majesty, 10th May, 1849*. London: W. Clowes and Sons, Stamford Street, for Her Majesty's Stationery Office, 1849.
3. Ewen, C.J. 1848. Journal. qMs-070S (ATL, NLNZ).
4. *New Zealand Spectator and Cook's Strait Guardian*, 6 November 1848.
5. *New Zealand Spectator and Cook's Strait Guardian*, 4 November 1848.
6. Mary Taylor to Ellen Nussey, 9 February 1849, in Berg Collection, New York Public Library. Reproduced in Joan Stevens (ed.) 1972. *Mary Taylor, Friend of Charlotte Bronte: Letters from New Zealand and Elsewhere*. Auckland University Press and Oxford University Press.

Chapter 2 A most painful duty

General: *New Zealand Spectator and Cook's Strait Guardian*, 18 October 1848; *The Wellington Independent*, 18 October 1848.

1. Lt. Gov. John Eyre to Governor George Grey, despatch, 19 October 1848, in *Papers relative to the Recent Earthquake at Wellington. Documents presented to both Houses of Parliament of Her Majesty, 10th May, 1849.* London: W. Clowes and Sons, Stamford Street, for Her Majesty's Stationery Office, 1849.

2. Ewen, C.J. 1848. Journal. qMs-070S (ATL, NLNZ).

3. Pratt, Rugby. 1877. *Colonial Experiences.* London: Chapman and Hall.

4. King, Thomas. Journal. King Family Papers MSX –4343 (ATL, NLNZ).

5. H.S. Chapman to H. Chapman, 18 October 1848, in Letters written between 1848 and 1849 by H.S. Chapman, Homewood, Karori, to his father, H. Chapman, 2 Tillotson Place, Waterloo Bridge, London. qMS-0419 (ATL, NLNZ)

6. Chapman, Judge Henry. 1849. Article in *Westminster and Foreign Quarterly Review*, Vol. 51, April–July.

7. McKain, Douglas. Transcript of journal entry by Robina McKain for October 1848 earthquakes. Micro-MS-0041 (ATL, NLNZ).

8. Oliver, Richard Aldworth. Private Journal 1848–9. Micro-MS-199 (ATL, NLNZ).

9. Ironside, Reverend Samuel. 1891. 'Missionary Reminiscences, No. XXIL, Story of the Earthquake Storms of 1848 – City in Ruins.' *The New Zealand Methodist*, 12 September.

Chapter 3 More of the same – only worse

General: *New Zealand Spectator and Cook's Strait Guardian*, 18, 25 and 28 October 1848; *The Wellington Independent*, 18, 25 and 28 October 1848.

1. Godley, John R. (ed.) 1951. *Letters from early New Zealand by Charlotte Godley 1850–1853.* Canterbury Centennial Edition. Christchurch: Whitcombe & Tombs Ltd.

2. Oliver, Richard Aldworth. Private Journal 1848–9. Micro-MS-199 (ATL, NLNZ).

3. Blake, A. Hope. 1909. *Sixty years in New Zealand: stories of peace and war.* Wellington: Gordon & Gotch Ltd, p. 43.

4. H.S. Chapman to H. Chapman, 18 October 1848, in Letters written between 1848 and 1849 by H.S. Chapman, Homewood, Karori, to his father, H. Chapman, 2 Tillotson Place, Waterloo Bridge, London. qMS-0419 (A-292-070) (ATL, NLNZ).

5. Chapman, Judge Henry. 1849. Article in *Westminster and Foreign Quarterly Review*, Vol. 51, April–July.

6. Young, J. 1901. *The Life of John Plimmer*. Wellington: N.Z. Times Co. Ltd. See also: 'Quakes in Wellington. The big shocks of '48 and '55.' *New Zealand Mail*, 9 May 1906, from 'Life of John Plimmer' of Wellington, a private publication, in which Plimmer refers to his experiences in Wellington during the 1848 and 1855 earthquakes.

7. Pratt, Rugby. 1877. *Colonial Experiences*. London: Chapman and Hall.

8. Ewen, C.J. 1848. Journal. qMs-070S; (ATL, NLNZ).

9. Ironside, Reverend Samuel. 1891. 'Missionary Reminiscences, No. XXIL, Story of the Earthquake Storms of 1848 – City in Ruins.' *The New Zealand Methodist*, 12 September.

10. *The Wellington Independent*, 25 October 1848.

11. *New Zealand Gazette* (Province of New Munster), No. 2, October 1848.

12. *The New Zealand Evangelist*, Vol. 1, No. V, 5 November 1848.

13. *The Wellington Independent*, 28 October 1848.

14. Lt. Gov. John Eyre to Governor George Grey, despatch, 19 October 1848, in *Papers relative to the Recent Earthquake at Wellington. Documents presented to both Houses of Parliament of Her Majesty, 10th May, 1849*. London: W. Clowes and Sons, Stamford Street, for Her Majesty's Stationery Office, 1849.

Chapter 4 The uncertainty continues

General: *New Zealand Spectator and Cook's Strait Guardian*, 25 and 28 October 1848; *The Wellington Independent*, 25 and 28 October 1848.

1. *The New Zealand Evangelist*, Vol. 1, No. V, 5 November 1848.

2. Ironside, Reverend Samuel. 1891. 'Missionary Reminiscences, No. XXIL, Story of the Earthquake Storms of 1848 – City in Ruins.' *The New Zealand Methodist*, 12 September.

3. Collinson, Thomas B. to his mother in England, 25 October 1848. MS Copy Micro 653, Letters 1846–49, Folders 1–3 (ATL, NLNZ).

4. McKain, Douglas. Transcript of journal entry by Robina McKain for October 1848 earthquakes. Micro-MS-0041 (ATL, NLNZ).

5. *New Zealand Spectator and Cook's Strait Guardian*, 8 and 15 November 1848.

6. Lt. Gov. John Eyre to H.E. The Governor in Chief, Despatch No. 116, Wellington, 20 November 1848. G-7, National Archives.

Chapter 5 North and East

General: *New Zealand Spectator and Cook's Strait Guardian*, 25 October, 1 November 1848; *The Wellington Independent*, 28 October, 1 and 8 November 1848.

1. Collinson, Thomas B. to his mother in England, 25 October 1848. MS Copy Micro 653, Letters 1846–49, Folders 1–3 (ATL, NLNZ).

2. Taylor, Reverend Richard. 1855. *Te Ika A Maui or New Zealand and its Inhabitants.* London: Wertheim and MacIntosh.

3. Taylor, Reverend Richard. Journal, Vol. 5, 12 March 1847–31 December 1848, qMS-1989 (ATL, NLNZ).

4. *The Wellington Independent*, 8 November 1848.

5. Oettli, Peter. 2008. *God's Messenger: J.F. Riemenschneider and Racial Conflict in 18th Century New Zealand.* Wellington: Huia Books.

6. Wilson, Dr Peter. 1849. First Annual Report, Colonial Hospital of New Plymouth. Taranaki Museum.

7. Chapman, Judge Henry. 1849. Article in *Westminster and Foreign Quarterly Review*, Vol. 51, April–July.

8. Hamlin, Reverend James. Letters and Journal of James Hamlin, 1835–1862. MS-0067-68, Hocken Library, University of Otago, Dunedin.

9. Groves, H.G. 1940. *Early Castle Point.* Groves Family Papers.

10. Colenso, William. Journal, Vol. 1, 1841–1848. qMS-0487 (ATL, NLNZ).

11. Horomona Pa, Turanganui, lower Wairarapa valley, to William Colenso, 20 October 1848, 91-169-1/8 (ATL, NLNZ).

12. Rihara Taki to William Colenso, 20 October 1848, Mitchell Library, New South Wales. From an extract translated and made available by Tahu Kukutai, St. John's Theological College of Te Pihoparanga o Aotearoa, Meadowbank, Auckland.

13. Taylor, Reverend Richard. Copy of a letter from Wanganui dated 18 November 1848, in Taylor Family Notebook No. 14, 1844–1873, MSX-5570 (ATL, NLNZ).

Chapter 6 The Middle Island

General: *The Nelson Examiner and New Zealand Chronicle*, 21 and 28 October, 11 November 1848; *New Zealand Spectator and Cook's Strait Guardian*, 25 October, 1 November, 6 December 1848; *The Wellington Independent*, 28 October, 1, 8, 22 and 25 November, 2, 6 and 9 December 1848.

1. Thomas Arnold to his mother, 10 October 1848, in Letters from New Zealand and Tasmania 1847–50. See also: Arnold, Thomas. 1900. *Passages in a Wandering Life.* London: Edward Arnold.

2. Manson, Celia. 1974. *The Story of a New Zealand Family, Part 1*. Queen Charlotte Sound: Cape Catley Ltd.

3. Gouland, Henry Godfrey. Diary 1801–1872. (ATL, NLNZ).

4. *The Wellington Independent*, 8 November 1848.

5. Constantine Augustus Dillon to his mother, Lady Dillon, 20 January 1849, in C.A. Sharp (ed.) 1954. *The Dillon Letters 1842–1853*. Wellington: A.H. & A.W. Reed.

6. Mary Frederica Swainson to her grandparents, 14 November 1848, number 54, qMS-1337-1339 (ATL, NLNZ).

7. Lyell, C. 1856. Folder entitled '1855 earthquake'. Gen 119, Special Collections, University of Edinburgh, Edinburgh, Scotland. Collection of miscellaneous papers that includes notes of an interview with Frederick Weld on 16 April 1856.

8. Saxton, John Waring. Diary 1841–1851. (ATL, NLNZ).

9. Janson, Hugh. 1967. 'Pioneer Memories at big Nelson Reunion,' *Evening Post*, 25 January (Memories of David Drummond).

10. Stephens, Samuel. Letters and Journals 1842–55, Nelson Provincial Museum. See also: Printed extract from *The Bristol* in Stephens, Samuel, MS-Papers – 2698-IA (ATL, NLNZ).

11. Alfred Domett to Superintendent of the Southern Division – New Munster (Major Matthew Richmond, Nelson), 25 October 1848. Letter No 247, SSD-1 (NA).

12. *The Wellington Independent*, 28 October 1848.

13. *The Wellington Independent*, 8 December 1848.

Chapter 7 Rap on the knuckles

1. Lt. Gov. John Eyre to Gov. George Grey, despatch, 19 October 1848, in *Papers relative to the Recent Earthquake at Wellington. Documents presented to both Houses of Parliament of Her Majesty, 10th May, 1849*. London: W. Clowes and Sons, Stamford Street, for Her Majesty's Stationery Office, 1849.

2. Manson, Cecil and Celia. 1962. 'Even Shaky Land Better than Cruel Sea', *The Dominion*, 20 July. Continued from an account in the 14 July issue.

3. *The New Zealander and Southern Cross*, 18 November 1848.

4. *The Nelson Examiner and New Zealand Chronicle*, 9 December 1848.

5. Godley, John R. (ed.) 1951. *Letters from early New Zealand by Charlotte Godley 1850–1853*. Canterbury Centennial Edition. Christchurch: Whitcombe & Tombs Ltd.

Chapter 8 Generosity declined with thanks

1. *The Wellington Independent*, 29 November 1848.
2. *The Wellington Independent*, 2 December 1848.

Chapter 9 A nine days' wonder

1. H.S. Chapman, Homewood, Karori, to his father, H. Chapman, 2 Tillotson Place, Waterloo Bridge, London, 1848–1849, qMS-0419 (ATL, NLNZ).
2. Fitzherbert, William. 1848. 'An account of the earthquakes in New Zealand.' *New South Wales Sporting and Literary Magazine and Racing Calendar*, Sydney: D. Wall, 76 York Street.
3. *New Zealand Evangelist*, Vol. 1, No. V, 5 November 1848.
4. *New Zealand Evangelist*, Vol. 1, No. VII, 1 January 1849.
5. Collinson, J.B., Park, R. & St. Hill, H. 1848. 'Report on earthquake damage in Wellington', *Government Gazette*, 6 December.
6. William Fox to the Secretary of the New Zealand Company, 31 October 1848, published in *The New Zealand Journal*, London, 19 May 1849.
7. Carter, Charles Rooking. 1849. Letter published in *The New Zealand Journal*, London, 19 May.

Chapter 10 Fire, water, air; a fissure in the ground

1. Fitzherbert, William. 1848. 'An account of the earthquakes in New Zealand', *New South Wales Sporting and Literary Magazine and Racing Calendar*. Sydney: D. Wall, 76 York Street.
2. Daubeny, Charles. 1848. *A description of active and extinct volcanoes, of earthquakes, and of thermal springs; with remarks on the causes of these phenomena, the character of their respective products, and their influence on the past and present condition of the Globe.* 2nd ed. London: Richard and John E. Taylor.
3. Field, H.C. 1898. 'Notes on the recent earthquake', *Transactions of the New Zealand Institute*.
4. Chapman, Judge Henry. 1849. Article in *Westminster and Foreign Quarterly Review*, Vol. 51, April–July.
5. *The Wellington Independent*, 25 November 1848.
6. *New Zealand Evangelist*, Vol. 1, No. V, 5 November 1848.
7. *New Zealand Evangelist*, Vol. 1, No. VI, ?? December 1848.

8. Thomas B. Collinson to his mother in England, 25 October 1848. MS Copy Micro 653, Letters 1846–49, Folders 1–3 (ATL, NLNZ).

9. *The Wellington Independent*, 28 October 1848.

10. Reverend Richard Taylor. Letter from Wanganui, 18 November 1848, MSX-5570 (ATL, NLNZ).

11. Strange, F. 1850. 'A narrative of a trip sixty-four miles to the west of Port Cooper.' *Sydney Morning Herald*, 26 January.

12. *Nelson Spectator*, 11 November 1848.

13. Saxton, John Waring. Diary 1841–1851 (ATL, NLNZ).

14. Weld, Frederick Aloyisius. 1848. Memoranda and Memorabilia, WELD 213, 4/1, 4/3, NA.

15. Stephen Nicholls to William Adams, 12 September 1851, in Journals of Martha and William Adams (1850–1852), MS-Copy-Micro-0344 (ATL, NLNZ).

16. Hochstetter, Ferdinand von. 1859. *Lecture on the Geology of the Province of Nelson.* Supplement to the *Nelson Examiner,* 1 October.

17. Lyell, Sir Charles. 1868. *Principles of Geology,* Vol. 2, 10th ed. London: John Murray.

18. Hochstetter, Ferdinand von. 1864. *Geology of New Zealand. Contributions to the geology of the provinces of Auckland and Nelson.* Translated from the German and edited by C.A. Fleming. Wellington: Government Printer, 1959.

19. Jones, Lieutenant Morton (HMS *Pandora*). Journal, Vol. 1. Mitchell Library, Sydney, qMS 1076 (ATL, NLNZ).

20. Jolliffe, John (surgeon on board HMS *Pandora*). The diary of John Jolliffe, R.N. New Zealand Journals 1851–1856. Mitchell Library, Sydney, Micro-MS-130, Reel 1 (ATL, NLNZ).

21. Musgrave, T. Survey map of January 1854 (S.O. 2941) and Survey Book. Department of Survey and Land Information, Blenheim.

22. Buckland, Reverend William. 1836. *Geology and Mineralogy.* London: William Pickering.

23. M. Richmond to Reverend Richard Taylor, 12 January 1849, in Sir George Grey New Zealand Manuscripts. Rev. R. Taylor Collection, Letters 1814–1873, Auckland Public Library, GNZ MSS 297/4.

Chapter 11 Interlude: searching for an earthquake and an important discovery

1. McKay, Alexander. 1888. Unpublished report on September 1st, 1888, and return journey from Wellington to Glenwye, Hope River, North Canterbury, between October

5 to October 17. MU 135, Folder 5, Box 3, 37 p. National Museum of New Zealand Te Papa Tongarewa.

2. McKay, Alexander. 1890. On the earthquakes of September 1888, in the Amuri District. *New Zealand Geological Survey Reports of Geological Explorations 1888–89*, 21: 1–16.

3. Freund, R. 1971. *The Hope Fault*. New Zealand Geological Survey Bulletin 86. New Zealand Department of Scientific and Industrial Research, 49 p.

4. McKay, Alexander. 1892. On the geology of Marlborough and south-east Nelson. [Part II]. *New Zealand Geological Survey Reports of Geological Explorations 1890–91*, 21: 1–28.

Chapter 12 The Awatere fissure: source of the 1848 earthquakes

1. Hector, J. 1890. Progress Report. Marlborough District (earthquake rents). *Colonial Museum and New Zealand Geological Survey Reports of Geological Explorations 1888–89*, 20: 36–53.

2. McKay, Alexander. 1889. Unpublished report, January 1st, 1889, on geological investigations in Awatere Valley. MU 135, Folder 5, Box 3, 37 p. National Museum of New Zealand Te Papa Tongarewa.

3. McKay, Alexander. 1902. *Report on the recent seismic disturbances within Cheviot County in Northern Canterbury and the Amuri District of Nelson, New Zealand (November and December 1901)*. Wellington: Government Printer, 472 p.

4. McKay, Alexander. 1890. On the geology of Marlborough and the Amuri District of Nelson. [Part I]. *New Zealand Geological Survey Reports of Geological Explorations 1888–89*, 20: 85–185.

5. McKay, Alexander 1892. *On the geology of Marlborough and south-east Nelson. [Part II]*. New Zealand Geological Survey Reports of Geological Explorations 1890-91, 21: 1-28.

6. Little, T.A., Grapes, R.H. & Berger, G.W. 1997. Late Quaternary strike-slip on the eastern part of the Awatere Fault, South Island, New Zealand. *Bulletin of the Geological Society of America*, 110: 127–148.

7. Grapes, R., Little, T. & Downes, G. 1998. Rupturing of the Awatere Fault during the 1848 October 16 Marlborough earthquake, New Zealand: historical and present day evidence. *New Zealand Journal of Geology and Geophysics*, 41: 387–399.

8. Benson, A.M., Little, T.A., Van Dissen, R.J., Hill, N. & Townsend, D.B. 2001. Late Quaternary paleoseismic history and surface rupture characteristics of the eastern

Awatere strike-slip fault. *Geological Society of America Bulletin,* 113: 1079–1091.

9. Mason D.P.M. & Little, T.A. 2006. Refined slip distribution and moment magnitude of the 1848 Marlborough earthquake, Awatere Fault, New Zealand. *New Zealand Journal of Geology and Geophysics,* 49: 375–382.

10. Cowan, H.A. 1991. The North Canterbury earthquake of September 1, 1888. *Journal of the Royal Society of New Zealand,* 21: 1–12.

11. Hill, N.H., Little, T.A. & Benson, A.M. 2001. *Paleoseismology of the eastern Awatere Fault since ~15 ka inferred from diatom biostratigraphy and sedimentology of Lake Jasper, New Zealand.* New Zealand Earthquake Commission Research Report 97/262, Chapter 5, 5.1–5.51.

12. Mason, D.P.M., Little, T.A. & Van Dissen, R.J. 2006. Refinements to the paleoseismic chronology of the eastern Awatere Fault from trenches near Upcot Saddle, Marlborough, New Zealand. *New Zealand Journal of Geology and Geophysics,* 49: 383–397.

14. Strange, F. 1850. 'A narrative of a trip sixty-four miles to the west of Port Cooper.' *Sydney Morning Herald,* 26 January.

15. Adams, C. Warren 1853. *A Spring in the Canterbury Settlement.* London: Longman, Brown, Green, and Longmans.

16. McDowell, James. 1910/11(?). Letter to the editor of the *Evening Post*(?), 9 August, in response to a lecture given by James McIntosh Bell, Director of the New Zealand Geological Survey, on Wellington Harbour and the earthquake risk to brick buildings in Wellington. Information in the letter is from the Family Scrapbook of Alistair Stuart of Gladstone, Wairarapa (Wairarapa Archives, Masterton, New Zealand). The information about ground damage between Port Robinson and Cheviot was related by Isaac Plimmer (son of John Plimmer, who experienced the 1848 earthquakes in Wellington).

17. Burrows, C.J. 1975. A 500-year-old landslide in the Acheron River valley, Canterbury. *New Zealand Journal of Geology and Geophysics,* 18: 357–360. See also: Cowan, H., Nicol, A. & Tonkin, P. 1996. A comparison of historical and paleoseismicity in a newly formed fault zone and a mature fault zone, North Canterbury, New Zealand. *Journal of Geophysical Research,* 101: 6021–6036.

18. Arnold, Thomas. 1900. *Passages in a Wandering Life.* London: Edward Arnold, pp. 97–99.

19. *New Zealand Spectator and Cook's Strait Guardian,* 15 November 1848. Extracts from the Meteorological Tables of Captain Oliver of the HMS *Fly* anchored in Lambton Harbour, Wellington.

20. Hochstetter, Ferdinand von. 1864. *Geology of New Zealand. Contributions to the geology of the provinces of Auckland and Nelson.* Translated from the German and edited by C.A. Fleming. Wellington: Government Printer, 1959.

Chapter 13 Future Visitation

1. Van Dissen, R., Taber, J.J., Stephenson, W.R., Sritharan, S., Read, S.A.L., Perris, N.D., McVerry, G.H., Heron, D.W., Hastie, W.J., Dellow, G.D., Campbell, H.J., Begg, J.G. & Barker, P.R. 1992. Assessment of seismic hazard in the Wellington region. *Recent Advances in Wellington Science,* Extended abstracts 8–9 July 1992, New Zealand Geological Survey Report G166, 49 p. + two maps.

2. Begg, J., Langridge, R., Van Dissen, R. & Little, T. 2008. *Wellington Fault: neotectonics and earthquake geology of the Wellington-Hutt Valley segment.* Geosciences '08 Field Trip Guides, Geological Society of New Zealand Miscellaneous Publication 124B, pp. 5–67.

MEMORIAL AND SCHEDULE OF LOSSES BY EARTHQUAKE, WELLINGTON 25TH JULY, 1849

Sir,

I have the honour to send herewith a Memorial of the inhabitants here who suffered by the recent Earthquakes, accompanied by a Schedule of the Losses sustained by each person, and a Government Gazette containing the Report of the Surveyors (dated the 21st of November last) showing the extent of damage done to the various buildings within the town, which I shall feel obliged by your Excellency forwarding to the Right Honourable Earl Grey Her Majesty's Principal Secretary of State for the Colonies; and I shall feel further obliged by your Excellency seconding the prayer of the memorialists for an application for a Parliamentary Grant for their relief. I have attached the Report of the Surveyors alluded to, because it gives a detailed official account of the disasters caused by the Earthquakes. In regard however to the estimated amount of damage in that Report, viz. £15,000, I may observe to your Excellency that that estimate is upon buildings only within the town; and does not include the losses of Merchandise and Household Goods which were to a considerable amount. I deem it necessary to make this observation to remove an apparent discrepancy that exists between the amount of losses stated in the Report, and that of the Memorialists in the Schedule; and which latter, to the best of my knowledge, I am convinced is within the actual amount of loss sustained.

A good deal of time has been unavoidably lost in drawing up the Memorial and getting the various signatures attached to it, and finding that it could not be despatched to your Excellency with a reasonable chance of its reaching England in time for the present sitting of the Imperial Parliament I have delayed forwarding it to your Excellency until the present period. A duplicate of the Memorial will be despatched to your Excellency by the next vessel sailing to Auckland.

I have the honour to be Sir,
Your most obedient servant,

Wm. Hickson.
Chairman of the Committee appointed to draw up and transmit the Memorial.

SCHEDULE OF BUILDINGS IN WELLINGTON AND THE NEIGHBOURHOOD WHICH HAVE BEEN DAMAGED BY THE EARTHQUAKES IN OCTOBER, 1848, REFERRED TO IN THE ACCOMPANYING REPORT

Situation	Property or Occupier	Nature of Building	Damage	Repairs Proposed by the Owner	Observations by the Board
Wellington Terrace	Mr. King, Solicitor	Clay house partly faced with brick; posts in the walls. The clay walls strengthened with slips of wood nailed across the posts about 9″ apart.	The brick facing down; the walls much shattered, part fallen down	To be pulled down and rebuilt of wood	None of these buildings are in a public thoroughfare and the Board do not think it necessary that they should offer any observations respecting them
	Mr. Cridland	One story house; walls of clay, with posts and slips nailed across	Walls shattered and partly down	Similar	
	Captain Sharpe	One story clay house; walls 12″ thick, with strong posts 2 feet apart; substantially built	The whole of the clay work much shaken	Being repaired with wood	
	Mr. Bethune	Two story brick house with verandah in front	Walls cracked	Under repair	
	Rev. J. O'Reilly	Two story clay house with thick walls; very well built	One of the gables much shaken, the other slightly; parts of the side wall loosened	Under repair	
	Mr. Strang	One story clay house faced with brickwork outside	Brickwork fallen out and the frony wall shattered	To repair with woodwork	

Situation	Property or Occupier	Nature of Building	Damage	Repairs Proposed by the Owner	Observations by the Board
Te Aro	Mr. Vincent	One story house part clay, part weatherboarded	One side wall of clay fallen out; one of the gables also	Rebuilt with wood	
	Mr. Plimmer	One story clay house with posts in the walls; well built; part faced with 4 ½" brickwork outside	Nearly all the brickwork down, and the walls shaken	Not known	
	Mr. Lowe	One story clay house; thick walls and well built	Both the gables down	Not known	
	Mrs. Hendry	One story clay house; walls 12" thick	All the clay work disturbed	Not Known	
	Mr. Hawkins	Similar	Part fallen; other parts falling	Not known	
	Mr. Forster	Similar	Front wall and one gable shattered; other parts loosened	Repaired with woodwork	
	Mr. Penny	Two story brick house with brick partitions; wood bonding all through at every 3 feet	Front and back walls thrust out; partitions split	Not known	
	Mr. Gooder	Small two story brick house	Completely shattered; partly down	Damages are being repaired in wood	
	Mr. Sutfield	One story brick house	Much shaken; wall cracked over the openings of windows, etc.	Not known	

Situation	Property or Occupier	Nature of Building	Damage	Repairs Proposed by the Owner	Observations by the Board
Te Aro	Mr. Masters	One story clay house	One side wall down; other parts shaken	Repaired with wood	
	Mr. Mudgrowy	One story clay house; posts in the walls and strips nailed across	Greater part down	Not known	
	Mr. Bennett	Similar	Similar	Not known	
	Mr. May	One story clay house; well built	Walls much rent	Not known	
	Mr. Ashdown	Similar	Slightly damaged	Not known	
	Mr. Mills	Similar	In ruins	To take down	
	Mr. Ford	Similar	Similar	To take down	
Sutton Row	Mr. Gerard	Well built clay houses; one story; small	One gable, and part of one side wall fallen out	Being repaired in wood	
		Several one story houses	All shattered; parts fallen down	To rebuild in wood	
	Military Hospital	Of clay; detached		To rebuild in wood	
	Mr. Quin	Several clay houses; one story; small	All more or less shaken; portions fallen down; all unsafe	To repair some with wood and take down others	
	Mr. Villars	Several small houses; some clay; others brick	Parts of the clay as well as the brick walls have fallen and whole have been much shattered	To repair some with wood and take down others	
Dixon Street	Mr. Howe	Two story brick house and wood bonding	One gable down; the rest slightly shaken	To repair with wood	
	Mr. Waters	One story clay house	One gable down	Repaired with woodwork	

Situation	Property or Occupier	Nature of Building	Damage	Repairs Proposed by the Owner	Observations by the Board
Dixon Street	Mr. Stoddard	One story; brick	Much shaken; one gable partly down	Repaired with woodwork	
	Mr. Blyth	Well built clay house; two stories with brickwork outside	All the walls shattered; portions thrown down	Being repaired with wood	
	Mr. Catchpool	Two story; brick. Flour mill	Shaken all over		Recommended to be taken down
Willis Street	Mr. Crowther	Two story; brick	Gables down; walls rent	Taken down	
	Mr. Wilkinson	Two story clay house	Front walls fallen out	Taken down	
		Several other clay and brick houses; small; one story	One entirely down; the others generally much shaken	Some repaired in woodwork; others taken down	
Manners Street		Wesleyan Chapel; large brick building	All down	Site being cleared	
	Mr. Rhodes	Store; two story; brick	Entirely down	Site being cleared	
	Mr. Hickson	Store; two story; brick	Both gables down	To rebuild in wood	
	Ridgeway's Co.	Large building	The side walls completely shattered	To rebuild in wood	
	Mr. Waitt	Store; wood with brick gable at S.E. end	Brick gable thrown out		Recommended that the bricks be cleared away

Situation	Property or Occupier	Nature of Building	Damage	Repairs Proposed by the Owner	Observations by the Board
Manners Street	Mr. Allen	Commercial room; one story; brick	Much shattered; part fallen down	To be boarded on the outside	Recommend to be done immediately; at present it endangers the public thoroughfare
		Public House; one story; brick and framing	Scarcely any damage	To be boarded on the outside	
	Messrs Bethune and Hunter	Store; framing, bricknogged and weatherboarding	Slightly damaged from the shifting of the packages inside	To repair	
	Mr. Fitzherbert	Stores; one story buildings. Framing bricknogged and brick boundary wall	The S.E. and N.W. ends thrown down and boundary wall also	To put a framework of wood inside and fix iron ties outside bolted to the framework	Recommended to be done immediately; the thoroughfare is dangerous at present
	H. M. Ordnance	Three story building; 13 ½" brickwork; no wood bonding; slate roof	Gables thrown out; N. wall cracked in several places	To be taken down	Recommended to be done immediately; the thoroughfare is dangerous at present
	Mr. Loxley	Store; brick	Gables much cracked	To be taken down and rebuild; gables in wood; put framework inside the walls with iron ties outside	Recommended to be done immediately; the thoroughfare is dangerous at present

Situation	Property or Occupier	Nature of Building	Damage	Repairs Proposed by the Owner	Observations by the Board
Manners Street	Union Bank of Australia	Weatherboard building; one story	The brickwork of the safe much shattered	To repair	
	Mr. Langdon	One story; brick lined with wood	Ends and side down	Rebuilding of wood	
	Mr. Hansard	Two story brick	Completely shattered	Taken down	

NEW ZEALAND EARTHQUAKE FELT INTENSITY SCALE

(from Geonet, GNS Science)

The size of an earthquake is often described using magnitude, which is the amount of energy released during an earthquake. However, depending on the earthquake's depth, not all of the energy released in an earthquake will necessarily be felt at the surface. In New Zealand, a felt intensity scale is a better indicator of an earthquake's effects on people and their environment. This was originally termed the Mercalli scale (named after the Italian volcanologist, Giuseppe Mercalli, who originally devised an intensity scale in 1884 and 1906) and the effects of earthquakes are graded into ten steps, using Roman numerals to distinguish them from the earthquake energy-release magnitude numbers. The scale used in New Zealand is ranked into twelve steps, hence it is termed *Modified Mercalli* (MM) scale, with I representing the weakest of shaking, through to XII representing almost complete destruction. The descriptions below are a simplified version of the New Zealand Modified Mercalli Intensity scale.

MM I: Imperceptible

Barely sensed only by a very few people.

MM II: Scarcely felt

Felt only by a few people at rest in houses or on upper floors.

MM III: Weak

Felt indoors as a light vibration. Hanging objects may swing slightly.

MM IV: Largely observed

Generally noticed indoors, but not outside, as a moderate vibration or jolt. Light sleepers may be awakened. Walls may creak, and glassware, crockery, doors or windows rattle.

MM V: Strong

Generally felt outside and by almost everyone indoors. Most sleeping persons are awakened and a few people alarmed. Small objects are shifted or overturned, and pictures knock against the wall. Some glassware and crockery may break, and loosely secured doors may swing open and shut.

MM VI: Slightly damaging

Felt by all. People and animals are alarmed, and many run outside. Walking steadily is difficult. Furniture and appliances may move on smooth surfaces, and objects fall from walls and shelves. Glassware and crockery break. Slight non-structural damage to buildings may occur.

MM VII: Damaging

General alarm. People experience difficulty standing. Furniture and appliances are shifted. Substantial damage to fragile or unsecured objects. A few weak buildings are damaged.

MM VIII: Heavily damaging

Alarm may approach panic. A few buildings are damaged and some weak buildings are destroyed.

MM IX: Destructive

Some buildings are damaged and many weak buildings are destroyed.

MM X: Very destructive

Many buildings are damaged and most weak buildings are destroyed.

MM XI: Devastating

Most buildings are damaged and many buildings are destroyed.

MM XII: Completely devastating

All buildings are damaged and most buildings are destroyed.

INDEX

65th Regiment 21, 23, 27, 31, 32, 44, 52

Acheron, HMS (sloop) 26, 39, 74, 116–117,
    120, 140, 149–150
Adams, Warren 150
aftershocks
    Christchurch (2010/2011) 13
    North Canterbury (1888) 127, 129, 136
    Wellington (1848) 41, 46, 48, 50, 51, 54–55,
        61, 72, 73, 77, 78, 100–101, 149, 150
    Wellington (1849) 109–110
    Wellington (future) 151, 154, 157
    see also under earthquakes
Aitkinson, William 120, 121
Allen, Mr 181
Alpine Fault **148**
Angus, George 113
Arnold, Thomas 67, **68**, 69–70, 71–73, 150
Ashdown, Mr 179
Auckland 84, 86, 89, 91–99
Aurora Australis 39, 41, 45–46, 77–78, 112, 116
Australian Plate **144**, 145, **148**
Awatere Fault **121**, 122, **123**, 124, **126**, 139,
    140, **141**, **142**, 143–147, **146**, **147**, **148**,
    162
Awatere fissure 118–122, **121**, **123**, 125–127,
    139–147, **140**, **142**
Awatere Valley **74**, 117–121, **121**, 122, 124–
    128, **126**, 127, 139, **140**, **141**, **142**

Baker, Major 15
Banks Peninsula 81, 116

Barefell Pass 119, 123, 127, 141, 145
barracks (Mount Cook) 33
barracks (Thorndon) **34**, 35
Barraud, Charles 17
Barrett Reef 18, **19**, 20
Bell, Francis Dillon 76
Bennett, Mr 179
Bernicia (ship) 81
Bethune, Mr 177, 181
Blyth, Mr 180
Board of Inquiry (1848 earthquakes) 103–104,
    177–182
Board of Inquiry (*Subraon*) 21
Brees, Samuel 34, 37, 46, 51, 60
Buckland, Professor Dean 121–122
building design 103–104, 106–109
building materials 177–182
    brick 29, 30, 32, 33, 35, 37, 41–42, 43, 55,
        56, 61, 71, 79, 80, 88, 103–104, 105–
        106, 108–109
    clay 29, 33, 43, 59, 61, 64, 70, 71, 79, 104,
        106, 108
    lime 104, 105–106, 108
    mortar 104, 108
    wood 27, 30, 55, 56, 84, 88, 103, 104, 130,
        153

Calder, James 17, 18–20, 21
Canterbury 13, 14, 124, **126**, 144, 148, 149,
    163
    *see also specific locations; specific geographic
        landmarks*

185

Cape Palliser 65

carbon dating *see* radiocarbon dating

Carpenter, Robert 96

Carter, Charles 107–109, **107**

Catchpool, Mr 180

Chaffers Passage 18, **19**

Chapman, Justice Henry 25–27, **28**, 30, 32, 33, 38, 41–42, 43, 61, 100–101, 109, 112–114, 116

Chile earthquake (1960) 9–10

Christchurch earthquakes (2010/2011) 13, 13n, 149, 163

Christchurch Fault (unnamed) **162**

Clara (ship) 29, 39

Clarence Fault 122, **126, 148**

Clay Point **17**, 24, **25**

Cloudy Bay 70, 72, 119, 122

Cole, Reverend Robert 45, 92–93, **93**, 99

Colenso, William 64–65, **65**, 66

Collinson, Captain Thomas 24, 55, 57, 59, 61–63, 103, 115

Colonial Government Barracks **60**, 62

Colonial Hospital 33, **40**, 41, 43

Conybeare, William 121–122

Cook Strait 78, 100, 120, 140, **156**, 158, 159, **162**, 163

Courts of Justice 32, **33**

Creed, Mr 36–37

Cridland, Mr 177

Crow, Captain 39–41

Crowther, Mr 180

Culverden 130

Dashwood Pass 120

Dillon, Constantine 71, **72**

Dixon Street 179–180

Domett, Alfred 67, 80

Dougherty family 70

Dougherty, Captain Daniel 72–73

Driver, HMS (sloop) 16

Drummond, Mr and Mrs 75

Earl Grey 83–85, 89–90, 175

earthquakes

    Chile (1960) 9–10

    Christchurch (2010/2011) 13, 13n, 149, 163

    epicentres 137–138, 144, 148

    fatalities 37–38, 42, 47–48, 50, 52, 158

    magnitude 9, 13, 144, 146, 148, 149, 155, 162, 163, 183–184

    North Canterbury (1888) 124–138, 142, 146

    psychological effects 11–12, 30, 38, 47, 49–50, 56, 72, 75, 77–78, 82, 101–103, 154, 161

    Wairarapa (1855) 13, 109, 139

    warnings 11, 23, 31, 32, 41

    Wellington (16th Oct 1848) 23–30, **40**, 57–58, 62, 68, 69, 73–77, 101, 139, 143–146, **144**, 149

    Wellington (17th Oct 1848) 31–39, **40**, 41, 77, 101

    Wellington (19th Oct 1848) **40**, 41, 42–48, 65, 77–78, 101

    Wellington (future) 12, 151–161, **162**

    see also aftershocks

epicentres 137–138, 144, 148

Episcopal Church 45

Ewen, Captain Charles 21, 23–24, 36, 38

Eyre, Lieutenant-Governor John 21, 23, 27, 32, 41, 43–44, 45, 51, 52, 53, 55–56, 82–83, **83**, 84, 85, 86–89, 90, 96, 103

fatalities 37–38, 42, 47–48, 50, 52, 158

faults 121–122, 147–149

    Alpine **148**

Awatere **121**, 122, **123**, 124, **126**, 139, **140**, **141**, **142**, 143–147, **146**, **147**, **148**, **162**
Christchurch (unnamed) **162**
Clarence 122, **126**, **148**
Greendale **148**, **162**
Hope 122, 124, 125, **126**, 130, **132**, **133**, 136, 137–138, 142, 146, **148**, **162**
Porter's Pass Fault 150, 150n
Wairarapa 122, **162**
Wairau **126**, **148**
Wellington 149, 159–160, **162**, 163
see also fissures
Featherston, Dr Isaac 95–96, 96–97, **97**
Felt Intensity Scale *see* Modified Mercalli Intensity Scale
ferry, interisland 158, 159
fissures 58–59, 69, 73, 111, 113, 114, 150
Awatere 118–122, **121**, **123**, 125–127, 139–147, **140**, **142**
Hope Fault 129, 131–132, **132**, 133, 134, 136–137,
Wellington 27, 63–64, 149, 159–160
see also faults
Fitzherbert, William 22, 181
Fitzherbert's store **40**, 41
Flaxbourne Station 67, **68**, 69–70, 150
flooding 41, 61, 158
Fly, HMS (sloop) 20, **24**, 29, 42, 55
Ford, Mr 179
Forster, Mr 178
Fort Richmond **46**
Fox, William 71, 97–98, 105, 106–107, **106**

geological mapping 121, 125, 126, 130, 131–134, **133**, 136–138
GeoNet 183
Gerard, Mr 179

Gladstone Station 120, 125–127
Glenwye 126, 131, **132**, **133**, 134
Glenwye Station 134–137, **135**, **136**
GNS Science 183
Gold, Colonel Charles **27**, 44, 102
Gooder, Mr 178
Government House **24**, 41, 43–44
Greendale Fault **148**, **162**
Grey, Sir George 23, 39, 82–85, **85**, 86, 87, 88–90, 93, 117

Hamlin, Reverend James 64
Hanmer Plain 124, 133
Hanmer Springs 128
Hansard, Dr James 22, 30, 182
Harlequin (schooner) 116
Havannah, HMS 85, 90, 93
Hawkins, Mr 178
Heale, Theophilus 91
Hector, Sir James 124, 143
Hendry, Mrs 178
Hickson and Co's store **40**, 41, 43
Hickson, William 52, 92, 175–176, 180
Hikurangi Pa 59
Hill, Henry St. 103
Hinchcliffe, Mr 52
Hochstetter, Ferdinand von 120, 141
Homewood (house of Henry Chapman) **28**
Hope
Fault 122, 124, 125, **126**, 130, **132**, **133**, 136, 137–138, 142, 146, **148**, **162**
River 133, 134
Valley 126, 134–137, **136**
Hort, Rabbi Abraham 93–94
Howe, Mr 179
Hunter, Mr 181
Hutt River 44, **46**

Hutt Valley 44–45, **46**, 104, 155, 156, 158, 159–160

Inglis, Reverend John 94–95
Ironside, Reverend Samuel 29, 36–37, 38, **39**, 52, 94, 98

Jacks, Mr 75
Jollie's Pass 127

Kaiparatihau River *see* Awatere River
Kaiparatihau *see* Awatere Valley
Kaitoke 159, 162, 163
Kanae (chief) 72
Kapiti Coast **60**
Kapiti Island **60**
Karori 25–27, **28**, 30, 38, 104, 154, 155, 159
King, Mary 25
King, Mr (solicitor) 177
King, Polly 25
King, Thomas 25

Lambton Harbour **16**, **17**, 18, **24**, **25**, **26**, 53–54, 157, 158, 159
see also Wellington Harbour
Lambton Quay 24, **25**, 36, 41, 44, **51**, 155, 157, 158
landslides 59, 102, 114, 120, 125, 129–130, 132–133, 134, 137, 149–150, 150n, 154, 156, 159, 163
Langdon, Mr 182
Lawry, Reverend Henry 91
Leslie Hills Station 130
liquefaction 60, 73, 128, 155, 159–160, 163
Lloyd, Mr 42
London (barque) **16**
Lovegrove, George 119–120

Lovell family 37–38, 42
Lovell, Sergeant James 37–38, 50, 52
Low, Mr 128
Lowe, Mr 178
Loxley, Mr 181
Lyell, Sir Charles 119, 122, 124
Lyttelton 81, 163

Magnitude Eight Plus (Grapes) 13
magnitude, earthquake 9, 13, 144, 146, 148, 149, 155, 162, 163, 183–184
Manners Street 30, **37**, **40**, 41, 157, 180–182
Maori 24, **25**, 57, 59, 65, 66, 70–71, 72, **73**, 81, 101
Marlborough 14, 67, **68**, 69–73, **74**, 145
see also specific locations; specific geographic landmarks
Mary Ann (ship) 80
Masters, Mr 179
May, Mr 179
McCleverty, Colonel 42, 52, 53
Mckay, Alexander 14, 122, 124–138, **125**, 139–145, 147
Memorial (Wellington 1848 earthquake) 175–176
Mercalli scale *see* Modified Mercalli Intensity Scale
Mercalli, Giuseppe 183
Methodist Church *see* Wesleyan Church
Military Hospital 33, 179
Mills, Captain J. P. 15–16, 18, 20, 21–22
Mills, Mr 179
Mills, Sergeant 103
Miramar 9, 155
Modified Mercalli Intensity Scale **144**, 183–184
Molesworth Station 127, 146
Montrose Station 130

Moore, George 15, 20
Mount Cook 24, **25**
Mount Victoria **24**
Mowat, Mrs 139
Mudgrowy, Mr 179
Mundy, Godfrey Charles 60
Musgrave, Thomas 120–121
Nelson 73–81, **76**
Nelson Examiner and New Zealand Chronicle
    73–75, 76, 79–80, 87
Nelson Spectator 118–119
New Zealand Company 18, 41, 95, 97, 105,
    106, 107, 117
New Zealand Editor and Southern Cross 86
New Zealand Evangelist 113, 115
New Zealand Journal 107
New Zealand Methodist 39
New Zealand Spectator and Cook's Strait
    Guardian 42, 64, 96, 149

Ngaio Gorge **102**
Nicholls, Stephen 120
Nicol, Janet 47–48
Nicol, Mr 47–48
North Canterbury earthquake (1888) 124–
    138, 142, 146

O'Reilly, Father Jeremy 45, 94, 177
Ohau River **60**
Oliver, Captain Richard 20, 29, 31, 42, 43, 47,
    53, 55
Operiki Pa 59
Ordnance store (H.M.) **40**, 41, 43, 181

pa 57, 114
    Hikurangi 59
    Operiki 59

Te Aro 24, **25**, 93
Tuamarina 72–73
Tuhitarata 65, 66
Waitohi 71
Pa, Horomona 65, 66
Pacific Plate **144**, 145, **148**
Paremata Point **60**
Park, Robert 22, 41, 103
Penny, Mr 178
Percival River 129
Petone 160
Petrel (cutter) 67, 70
Petrie, Mrs 31–32
Picton 71, 74
Pipitea Point **17**
Pipitea Street **40**, 41
plate tectonics 144–149, **144**, **148**
Plimmer, John **34**, 35, 154, 178
Porirua 104
Porirua Harbour **60**
Port Cooper *see* Lyttelton
Port Nicholson *see* Wellington Harbour
Porter's Pass Fault 150, 150n
Pratt, Rugby 24, 29–30, 35–36, 44
psychological effects (of earthquakes) 11–12,
    30, 38, 47, 49–50, 56, 72, 75, 77–78, 82,
    101–103, 154, 161

Queen Charlotte Sound 71
Quin, Mr 179

radiocarbon dating 147, 150n, 163
religion 12, 18, 39, 44–45, 49–50, 52–53,
    92–93, 101–103
rents *see* fissures
Rhodes, W. B. 15, 21, 35, 43, 52, 180
Richmond, George 92

Richmond, Major Matthew 80, 118–119, **118**

Richter Magnitude Scale 9, 13, 144, 146, 148, 149, 155, 162, 163, 183–184

Ridgeway's Company 180

Riemenschneider, William 61

Riwaka 75

Rongotai Isthmus 9, 155, 163

Roots, Joseph 95, 96

Rotherham 130

rupture *see* faults; fissures

Rutherford, Duncan 130

Sarah Anne (ship) 29, 77

Saxton, John 75, **76**

Schedule of Buildings Damaged, 34, 177–182

Scotch Church 44, **51**

seiching 163

   see also tsunami

seismology 111–112, 114–117, 124–138, 139–150, 160

Selwyn, Bishop George 91, **92**

Sharpe, Captain 21, 177

slips *see* landslides

St Helens 128, 130

Stephens, Samuel 77, 78, 80

Stoddard, Mr 180

Stokes, Robert 96

Strang, Mr 177

Strange, Frederick 116–117, 149–150, 150n

Subraon (barque) 15, **16**, 18–22, **19**, 45, 47, 52

Sutfield, Mr 178

Sutton Row 179

Taylor, Reverend Richard 59, 60, 61, 63–64, 112, 116

Te Aro **17**, 24, **25**, 30, 32–33, 35, **37**, **40**, 41–42, 44, 61, 101, 155, 157, 178–179

Te Aro Pa 24, **25**, 93

Te Heu Heu (chief) 113, 114

Te Rauparaha (chief) 69, 72, **73**

Terrace, The 177

Thompson, Mr 136

Thorndon **17**, **24**, **25**, 31–32, 33–35, **33**, **34**, **40**, 41, 101, 149, 155

tidal movement 41, 61, 158

Torlesse Range 149–150, 150n

tsunami 61

   Chilean earthquake (1960) 9–10

   Wellington earthquake (future) 158, 163

Tuamarina 72

Tuamarina Pa 72–73

Tuhitarata Pa 65, 66

Turanganui 65, 66

Union Bank of Australia 182

Upcot 120, 125–127

Vernoon Lagoons **70–71**

Villars, Mr 179

Vincent, Mr 178

volcanoes 46, 112, 114, 115, 116

   Mt Etna 111

   Ngauruhoe 112, **113**

   Ruapehu **113**

   Tongariro 41, **113**, 114, 115

Waiau River 129–130, 131, 133, 134

Waimate 61

Wairarapa 13, 65

earthquake (1855) 13, 109, 139

Fault 122, **162**

Wairau

   Fault **126**, **148**

   Massacre 72

Plain **70–71**

Purchase 117

River **70–71**, 73

Valley 72, 117–118, 120

Waitangi (Hawkes Bay) 64

Waitohi 71, 74

Waitohi Pa 71

Waitt, Robert 51, 180

Wanganui 57, 58, **59**, 61, 63

Wanganui River 57, 58–59, **59**

Warea 61

Waters, Mr 179

Weld, Frederick 67, 68, 69–70, 71–73, 119–
    120, **119**, 122, 123, 125, 150

Wellington 17–18, **17**, **24**, **25**, **26**, **102**
    *see also specific locations; specific geographic
        landmarks*

Wellington Airport 157, 158, 163

Wellington earthquakes
    1848 (16th Oct) 23–30, **40**, 57–58, 62, 68,
        69, 73–77, 101, 139, 143–146, **144**, 149
    1848 (17th Oct) 31–39, **40**, 41, 77, 101
    1848 (19th Oct) **40**, 41, 42–48, 65, 77–78,
        101
    future 12, 151–161, **162**

Wellington Fault 149, 159–160, **162**, 163

Wellington Harbour **16**, 18, **19**, **24**, **25**, **26**,
    53–54, **156**, 159, 163

Wellington Independent 52–53, 53–54, 61, 65,
    72, 80–81, 99, 101, 114

Wesleyan Church 29, 33, 36–37, **37**, 43

White Bluffs 71, 118, 119, 120, 122, 123, **140**,
    141, 145, 146

White Island 113

Wilkinson, Mr 180

Willis Street 157, 180

Wilson, Dr Peter 61

Woodward, Jonas 95, 98

Woon, Reverend 61